The Pristine Dao

SUNY series in

CHINESE PHILOSOPHY AND CULTURE

Roger T. Ames, *editor*

The Pristine Dao

Metaphysics in Early Daoist Discourse

Thomas Michael

State University of New York Press

Published by
State University of New York Press, Albany

Printed in the United States of America

For information, address State University of New York Press,
194 Washington Avenue, Suite 305, Albany, NY 12210-2365

Production, Laurie D. Searl
Marketing, Susan Petrie

Library of Congress Cataloging-in-Publication Data

Michael, Thomas, 1966–
The pristine Dao : metaphysics in early Daoist discourse / Thomas Michael.
p. cm. — (SUNY series in Chinese philosophy and culture)
Includes bibliographical references and index.
ISBN 0-7914-6475-X (alk. paper) — ISBN 0-7914-6476-8 (pbk. : alk. paper)
1. Taoism. 2. Philosophy, Taoist. 3. Cosmogony, Ancient. I. Title. II. Series.
BL1920.M53 2005
181′.114 — dc22

2004017949

10 9 8 7 6 5 4 3 2 1

for Patricia Roberts and Keith Bishop

Contents

Acknowledgments

I had been studying in college for a year and a half when I decided to drop out, and ended up in India for six months. After falling in love with Hindu culture, and realizing that I was out of money, I decided to go back to college with the intent to study Sanskrit. My college did not offer any Sanskrit or Hindi courses, so my next safest bet to study something that fed my appetite for the alien was Chinese. I took a course on Chinese Literature in Translation, and ended up writing my term paper on Chinese fishermen. Most of them happened to be Daoist. Thus began my fascination with the materials discussed herein.

Although I think that an author should assume responsibility for the rights and wrongs of his or her book, some of the "rights" of the current work have been out of my control. Other responsible parties need to take the responsibility for them. Jonathan Pease first taught me about those fishermen. Rita Vistica went out of her way to make sure I did not drop out of college a second time. Odun Arechaga opened many important doors for me. Anthony Yu and Wendy Doniger assisted in the laborious delivery of the book. Sarengaowa helped the book to speak Chinese. Jane Geaney helped it to speak English. Henry Rosemont, Jr., introduced it to the world. And Nancy Ellegate made sure it would stay there. My deepest thanks to all of you.

I want to extend a special note of gratitude to the following people for their support that allowed me to push on with this book: Simone Krause, Hiroyoshi Noto, Jim Fitzgerald, Paul Duff, and the anonymous reader at the State University of New York Press. Again, my deepest thanks.

Note on the Texts

When they were available, I have relied on the standard *Sibu Beiyao* editions for all of my translations and citations. Because some of the texts that I examine have been excavated only very recently, I have relied on the best modern editions available. In the Bibliography, I name the primary sources with the edition, their standard dates as established by modern scholarship, and the abbreviations that I use in each citation.

Chapter One

Early Daoism and Metaphysics

Despite the scholarship of the last thirty years, early Daoism is still a controversial issue. The controversy centers on the religious nature of Chinese Daoism as a whole: does it become a religious tradition only with the revelation of the deity Laojun 老君 to Zhang Daoling 張道陵 in 142 CE, or is there some connection between this institutionalized Daoist religion and what went on in the textualized ideas that circulated for three, four, or even five hundred years before that?

Any reader already familiar with Daoism will be struck by the problem that leaps out not only from the first line of this work, but also from the general framing of the project represented herein. I allude to something called "early Daoism," a term that will undoubtedly surprise and quite possibly irritate not a few readers. The standard view of the history of early Daoism commonly posits a strict separation between, on the one hand, the early Daoist philosophical texts of the *Laozi* 老子 and the *Zhuangzi* 莊子, collectively represented in the modern academy by the term *daojia* 道家 or "Daoist philosophy," and, on the other hand, the later Daoist religious institutions that historically appeared after 142 CE, represented by the term *daojiao* 道教 or "Daoist religion." One motivation of this project, possibly its main one, is to attempt to wipe away the dichotomy imagined in many quarters between "philosophical" and "religious" Daoism. While the issue may not have everything to do with the particulars of what is intended to be signified by use of the label "Daoism," the quandary remains: how are we to avoid the general tendency of referring to "traditions" by these "ism" labels and the consequent characterizations necessarily leading into essentialized assumptions of such entities and definitions of their boundaries vis-à-vis other such entities? Although there is a certain heuristic value in doing so, it also limits and falsifies in other ways. This essentializing habit, employed by most people who speak and write, is for all of that still grounded in metaphors.[1] The nominalist stand against this habit would claim that such entities do not exist outside of some people's constructions undertaken for certain purposes; what does exist are people and the writings they produced and the cultural objects that they made.

None of this is to argue that there is not a demonstrable thematic, verbal, and intellectual cohesion among the writings that I term "early Daoist"; that demonstration, indeed, is one of the main goals of this project. In making this claim, I am not arguing for the de facto existence of an institutionalized Daoist community before 142 CE, nor will I provide a systematic theology of Daoist thought. I have consciously tried to refrain from making any claims, one way or the other, concerning any kind of early Chinese social group that may or may not be called Daoist, and I cannot emphasize this point enough. Leaving aside those sociological arguments, I do find the undeniable presence of a discursive tradition that undeniably existed before the Han dynasty. Since ideas have legs and do not float on thin air, who these people were that kept the discourse alive is open to speculation, but it is not the concern of this work. The meaningful referent for what I here call early Daoism is, in the most general way, not more than a discursive tradition. This discourse is infused with a distinct and coherent intelligibility that is best analyzed by examining its presentation within the corpus of writings that can be shown to participate in it.

This work confronts that intelligibility and, by identifying the complex of shared ideas found in those writings that participate in this early Daoist discourse but not found in other writings that do not, offers a new way to read what I call early Daoist discourse. Even though these ideas do not call out "I'm Daoist!" (if they did, there would be no need for this project), the presence of that thematic, verbal, and intellectual cohesion is fairly clear once we move beyond the constraining impositions of locating this tradition as either religious or philosophical. These ideas, nonetheless, still clearly participated in the general movement of religious and philosophical thought and belief in early China; more specifically, the writings that I call early Daoist are most readily identified through their participation in a common complex of notions about the pristine Dao not shared by any other writings from any other traditions. I recognize any piece of writing as early Daoist if it demonstrates an active participation in this complex of notions cohering around specific conceptions of the pristine Dao. This complex of notions not only gives a distinctive character to early Daoist disputation of the Dao but can also be shown to share certain fundamental assumptions with later, organized Daoism.

The complex of notions located in the early Daoist writings taken all together constitutes what I call early Daoist discourse. I argue that the prominent themes of this discourse are sustained and given coherence in four primary domains: cosmogony, cosmology, ontology, and soteriology. In distinguishing these domains, we see a discernable profile to the early Daoist discourse that is anything but random and that also continues to be recognized in the writings of later Daoism through their active adoption of the terms, images, and themes established by that discourse. Rather than setting forth in this introduction the identifying elements of this discourse point by point as they are found in discrete passages in the writings under consideration, I attempt, as much as possible, to allow the writings to speak for themselves while remaining aware of the violence I inevitably perform on them by cutting them up and placing them in the set of categories that I take from my own, non-early Chinese intellectual tradition.

Chronologically, "early Daoism" signifies the interval material that began to appear from the time of the traditional positing of the man Laozi at the end of the Spring and Autumn period, or at least from the first appearance of the *Laozi* writings, the date of which is generally established either by the Guodian edition of the *Laozi* or the earliest period of the Jixia academy, dates very close to those of Confucius.[2] The period of early Daoism comes to a close abruptly with the death of the prince of Huainan 淮南子 in 122 BCE. Huainan's death by execution was symbolic of the dramatic events that witnessed the elevation of the Confucian ideology to the level of state orthodoxy at the hands of Emperor Wu 漢武帝, Sima Qian 司馬遷, and Dong Zhongshu 董仲舒. The examination of the historical records giving information about the next 250 years of Daoist history, from the death of Huainan to the revelation of Zhang Daoling, remains a separate project that lies beyond the limits of this work.

Previous attempts by modern scholars to deploy some sort of interpretative framework onto the writings of early Daoism commonly have led to the imposition of distinctions that are completely foreign not only to the spirit of early Daoism but to early Chinese thought and practice as a whole. In the 1950s, Herlee Creel was the first Western scholar to apply a sophisticated interpretative model to early Daoism. He designated two different kinds of Daoism, the first of which he called contemplative and the second purposive. "Contemplative Daoism" is characterized by what he called "a mad intoxication with the wonder and power of nature," and is best exemplified by the *Zhuangzi*.[3] "Purposive Daoism" is characterized by "the attempt to utilize an essentially mystical doctrine for the furtherment of personal ambitions and political purposes," and is best exemplified by the *Laozi*.[4] Together, these two forms of early Daoism were to be known as "philosophical Daoism," in distinction to *xian* Daoism that is based on the quest for immortality.[5] The lasting consequence of Creel's model for later interpretations of early Daoism was to draw a clear line of demarcation between, on the one hand, the thought of the *Laozi* and the thought of the *Zhuangzi*, and, on the other hand, the ideas they embodied in distinction to all other Daoist writngs.

Maintaining this clear distinction between the *Laozi* and the *Zhuangzi*, A. C. Graham some thirty years later presented a new model for interpreting early Daoist writings based on the distinctions he drew between what he saw as the different voices in the *Zhuangzi* itself with the specific chapters attributable to each.[6] He categorized these as the Individualist, the Primitivist, and the Syncretist. This breakdown was extended by both Graham himself and other scholars to refer not only to the different voices in the *Zhuangzi*, but also to all of the other writings of early Daoism. For him, the Individualist writings emphasize spontaneity, mysticism, and self-cultivation, and are best exemplified in the first seven chapters of the *Zhuangzi*. The Primitivist is similar to the Individualist, but adds to its topics of attention a cosmological theory together with a specific body of political theory, and is best exemplified by the *Laozi*. Last is the Syncretist layer, which Graham characterizes by its assimilation of the thinking and goals of the several other early Chinese intellectual traditions. It is directed specifically to the ruler, and gives a strong priority to government rather than to individual self-cultivation; for him, this category is best exemplified by the *Huainanzi*. It is interesting to note that Graham retained, albeit in a slightly revised form, Creel's basic set of distinctions between the *Laozi* and the *Zhuangzi*. By using this model, Graham is able to designate discrete portions of the early Daoist writings as belonging to any one of these three groupings, thereby thwarting possible notions pertaining to the internal coherence of the early Daoist discursive tradition as a whole.[7]

Directly drawing on the kind of distinctions set forth by Graham, Harold Roth has presented the most recent hermeneutical model applied to the writings of early Daoism. He interprets them through three general categories of specifically Daoist concern:

1. Cosmology: a cosmology based on the Dao as the predominant unifying power in the cosmos.
2. Inner cultivation: the attainment of the Dao through a process of emptying out the usual contents of the conscious mind until a profound experience of tranquility is attained.
3. Political thought: the application of this cosmology and this method of self-cultivation to the problems of rulership.[8]

Roth's interpretative framework somewhat closes the gap between the separations inserted into the early Daoist tradition by writers like Creel and Graham. Furthermore, by relying on this framework, he is able to bring into the category of early Daoism many writings that scholars traditionally have not recognized as such; according to Roth, if a discrete piece of early Chinese writing meets the criteria of what counts as belonging to one of these three categories, it is to be identified as early Daoism.

Roth's categories, however, insert a different set of distinctions onto the integrity of early Daoist writings, of which the most significant is the distinction between body and mind or spirit. He distinguishes the categories of "cosmology" and "inner cultivation" in order to demonstrate the early Daoist concern with identifying with the pristine Dao through a kind of mysticism. In other words, the separation between what he understands as "cosmology" and "inner cultivation" is overcome by a mystic transcendence of the human body as the spirit identifies with the Dao. In part, Roth's distinction between "cosmology" and "inner cultivation" is the result of too strict a separation between a material cosmology involving a cosmic Dao and a mystical self-cultivation involving an ineffable Dao. The Dao of the cosmology and the Dao to which a person unites are more closely related than he presents them: both are fundamentally grounded in a physicality that can never be brushed aside. Furthermore, Roth's third category, "political thought," refers to the material consequences for the world that follow from the mystical attainment of an individual, either a sage or a ruler. This third category marks his interpretative effort to close the space between the mystical spirit and the material body that his categories have created. However, the early Daoist writings do not privilege the political as an independent realm of discourse; rather, the political benefits for the country accrue from the ability of a sage, as one who has made the Dao fully present in the world by embodiment, to act as a kind of conduit whereby the Dao can easily access the human world that human beings have caused to be closed off from it. Thus, although his categories partially correct those of Creel and Graham, the distinctions Roth draws are grounded in a metaphysical split that cannot be sustained by any reading of the writings themselves.

Categories such as these are the product of modern scholarly attempts to provide some type of hermeneutical access to writings traditionally recognized as Daoist. Although the writings do not employ anything like the kinds of categories presented by Creel, Graham, and Roth, there is still no way to avoid the need to apply some set of categories to the materials in order to generate intelligibility for us, and this work represents no exception. My choices for the categories employed in this project owe much to the work of earlier scholars like Creel, Graham, and Roth, but I hope to overcome some of the more problematic issues that their categories have created. The central categories—or, as I prefer to call them, domains of discourse—that I employ to make possible my own access to the ideas in the writings emerge from the vocabulary of metaphysics and theology. These domains include cosmogony (chapter 2), cosmology (chapter 3), ontology (chapter 4), and soteriology (chapter 5). By employing this terminology, I do not mean that all Daoist writings have a cosmogony, cosmology, ontology, and soteriology, nor that they even have these kinds of categories available to their own systems of thought (though this would be hard to prove or disprove). All the writings I consider address a lot of questions, all at once, and I will examine these questions in terms of the four domains, one by one in each of the remaining chapters.

The kinds of distinctions that necessarily must be applied in isolating these domains inevitably falsify; I do not intend to categorize individual writings standing alone as representing one domain nor that all of the domains are found in any single writing. I will call upon these domains only so far as they can assist, in a general way, the intelligibility of my readings of discrete pieces of writing, or sustained sections, within the substantial collection of surviving writings of early China, only some of which can be identified as specifically Daoist. Other early Chinese traditions also discuss and explore these four domains of discourse but, I will argue, only coincidentally and in piecemeal fashion; only early Daoist writings construct a complete, sustained, and coherent vision of reality and experience that can be interpreted as oriented around these four domains.

The first three of these domains (cosmogony, cosmology, and ontology) address the three central questions of metaphysics, and the fourth (soteriology) addresses the central question of theology not dealt with by metaphysics. Metaphysics begins with the questions of ontology: "What is there in the universe?" (Minds? Bodies? Stuff? Ghosts? Spirits? Angels?). It then asks the questions of cosmogony: "Whatever there is in the universe, how did it originate?" (Genesis? Brahma? Shunyata?). It finally asks the questions of cosmology: "Whatever there is in the universe, how do the pieces of it relate to each other?" (Mind-body problem, how many angels dance on the head of a pin, reductionism). Theology adds a further consideration with its questions of soteriology: "Whatever there is in the universe, where does it lead?" (Salvation or damnation? Utopia? Democracy, theocracy, or socialism?).

Despite their apparent sequential progression, in this work I start with cosmogony, not ontology. This should come as no surprise because the mechanisms of early Chinese cosmology and ontology, as well as cosmology and soteriology, revolve around the temporally constant movements of the pristine Dao. In the early Daoist writings, questions about the pristine Dao in and of itself, before and outside of the world, have a certain primacy that make the questions of cosmology, ontology, and soteriology understandable. To call upon a simple cliché, time and space in early China tend more toward cyclicity than unilinearity, and since each of the domains I discuss constantly returns to all of the others without leaving them behind once and for all, why not start at the beginning?

Chapter Two

Early Daoism and Cosmogony

BEFORE THE WORLD

The cosmogonies of early Daoist discourse are characterized by sustained explorations of the formation of time and space in relation to the prior existence of the pristine Dao. Outside of the early Daoist writings, various other early Chinese discourses also explore cosmogonic beginnings, but they are limited in scope and not systematic. In Mircea Eliade's phrase, they do not reach back to events *illore tempore*, the time before time that earlier human societies creatively attempted to represent in their myths of origins.[1] The images and motifs employed in the mythological visions of the very origins of things serve to make that time more real, more accessible to the mind and the imagination. The early Chinese, before the appearance of early Daoist discourse, richly employ the images and motifs from their own myths in depicting the beginnings of things, but they lack a clear notion of a time before time, before the world existed in *illo tempore*. One of the main reasons the writers of early Daoist discourse could explore and express the very origins in their speculations is that they develop a notion of a pristine source, which they named the Dao, and around which they frame a complete and coherent cosmogony for the first time in the historical records of early China. In doing so, however, they rely heavily on the images and motifs of the more generally disseminated body of preexisting Chinese myth. In the steady construction of their cosmogonies, these writers consistently employ those same images and motifs, but they are no longer primary. Instead they become secondary materializations of the separate components of the pristine Dao, and are used to represent the processes and sequences of the unfolding of the Dao in time and space.

The images and motifs employed in early Daoist writings are taken almost exclusively from the mythologies of the southern cultures. In the present time, our knowledge of these images and motifs is rapidly expanding due to the wealth of sites being unearthed from the regions of the early southern cultures, especially Chu. Although the images and motifs bear an unmistakable association with themes of the spatiality and orientation of the physical world, they in no way represent the time before the primordial coming-to-be of the world.[2] For the moment, I want to examine in some detail the motif of the dragon-tiger pair and its probable position within early Chinese myth. One early example of this dragon-tiger pair is found on the covers of two boxes from the excavated tomb of Zeng Hou Yi from Suixian, located in what was the Chu heartland, and dated to the late fifth century. In her discussion of these Chu objects, Jenny So writes,

> On the lid of one box are written the names of the twenty-eight lunar lodges, surrounding a large central depiction of the Northern Dipper constellation and flanked by a dragon and tiger, the animals of the east and west. The cover of the second box shows, along one edge, intertwined serpents with human heads, a motif that, by Han period, became the standard image for the gods Fu Xi and Nu Gua, the progenitors of the universe in Chinese myth.[3]

Leaving aside the later myths of Fu Xi 伏戲 and Nu Gua 女媧, here I want to examine briefly the mythological imagery that associates dragons, snakes, watery chaos, and generation through the incestuous intercourse of a primordial couple. Although explicit identifications of this primordial couple as Fu Xi and Nu Gua are rare before the Han (but commonplace thereafter), images of intertwined serpentine, dragonlike figures are prevalent in the archaeological remains from the southern domains of early China. According to Norman Girardot, "The uniped and serpent qualities of these figures, especially Nu Gua, are significant in that suggestions of a dragon nature are frequently linked with the mythic figures in the chaos theme and are central to the southern tribal mythology."[4] More than representing simple portrayals of artistic motifs, dragons also are carriers of an abundance of symbolic significance, as Mark Lewis points out.

> Many of the achievements of Fu Xi and Nu Gua are directly linked to the characteristics of the dragon in early Chinese thought. First, the movements of the dragon make it a link between Heaven and Earth. Through its association with water, the dragon could either hide in the depths of rivers or oceans or soar up on the clouds. This ability to move between Earth and Heaven enabled dragons to act as chariot steeds for those who set out on spirit journeys.[5]

Images of snake-dragon animals and the shamans who wore them as ornaments of power are legion in the *Shanhaijing* 山海經 and the *Chuci* 楚辭, which provide many images as well of the relation between dragons, shamans, and ecstatic flight. The famous Chu silk manuscript, dating from roughly 300 BCE, relates the earliest recorded myth concerning Fu Xi and his female counterpart, here named Nu Tian. This text provides a depiction of the earliest times recorded by the Chu peoples and the role played by Fu Xi and his female counterpart in the establishment of the physical world from the throes of a primordial chaos.

> Long, long ago, Bao Xi [= Fu Xi] of [. . .] came from [. . .] and lived in [. . .]. His [. . .] was [. . .] and [. . .] woman. It was confusing and dark, without [. . .], [. . .] water [. . .] wind and rain were thus obstructed. He then married Zuwei, [. . .]'s granddaughter, named Nu Tian. She gave birth to four [. . . (children)] who then helped put things in motion, making the transformations arrive according (to Heaven's plan). Relinquishing (this) duty, they then rested and acted (in turn) controlling the sidewalls (of the calendrical plan); they helped calculate time by steps. They separated (Heaven) above and (Earth) below. Since the mountains were out of order, they then named the mountains, rivers, and four seas. They arranged (themselves) by [. . .] hot and cold *qi*. In order to cross mountains, rivers, and streams (of various types) when there was as yet no sun or moon (for a guide), when the people traveled across mountains and rivers, the four gods stepped in succession to indicate the year; these are the four seasons.[6]

This complex of mythological symbols and motifs, including the primordial couple, dragon-snakes, and a watery chaos, functions as markers of sense signification in the depictions of the earliest formation of the world. These markers, proceeding by way of binary pairs (dragon-snake and male-female) are given more manifest form in the figures of Fu Xi and Nu Tian. Together they act upon the already existing chaotic mass of the primordial world, and carve out the fundamental features of the physical world, making it secure and habitable for human beings. The Chu silk manuscript appears representative of the various records of mythological speculation preceding the historical emergence of early Daoist discourse, in that there is no demonstrable awareness of a time preceding the existence of the chaotic mass: the world is always already there. There are no other archaeological records that show the origins of the primordial couple; all that can be said is that with the stories depicting their mythological activities, the world begins to be formed as a place habitable by human beings.

The text also makes mention of the "hot and cold *qi*," which appears for all intents to come after the existence of the primordial world. By contrast, when early Daoist writings sometimes mention "hot and cold *qi*" (and the *Taiyi Sheng Shui* serves as an early example of this), this usage serves as a byword for discussions concerning *yin-yang*. In the hands of the early Daoist writers, *yin-yang* typically precede the existence of the world, and stand as the generative source for its gradual formation from the body and movements of the pristine Dao. The development of the early Daoist notions concerning *yin-yang* owed a great deal to the ancient mythological images of dragons, snakes, and other primordial beings that were assimilated into them, and are sometimes applied as metaphors for the invisible processes identified with *yin-yang*. For early Daoist writers, depictions of the generation of the world through the sexuality of a primordial couple take on greater sophistication because of their developed theory of the Dao, making possible their discussions of the time before time. Thus, the early Daoist writings are able to explain the first origins of the world as a process of the unfolding of the Dao, and motifs of the primordial couple are applied as metaphors for *yin-yang*.

There are numerous interesting instances in which the early Daoist writings call upon these mythological images and their symbolic attributes in the discursive production of their own depictions of the beginnings. A good example comes from the opening sections of *Zhuangzi* 1, which provides evidence of the subtle transformation of these images and motifs in their application within a specifically Daoist depiction of the earliest beginnings. Notice the extremely sophisticated nature of this passage; instead of a naive presentation of a basic myth, *Zhuangzi* 1 offers three variations of the same material. The three variations consist of, first, a neutral and objective depiction of the brute events of the beginning; second, the same events written into a textual or historical record; and, third, as the topic of a debate between a king and a minister. This three-part narrative structure allows the writer to express the radical separation between the time of the earliest beginnings and the present age, thereby indicating the inherent inability for human beings living in the present age to represent and thus have firsthand knowledge of those beginnings. In doing this, the writer arrives at an even more fundamental object of knowledge—namely, the cyclic nature of the binary forces of *yin-yang* as cosmogonic elements and their continuing effects on the lived world. By way of an initial reliance on the mythological images, the writer in effect subsumes them to his own discourse in presenting a cosmogonic vision demonstrating some fairly typical characteristics of early Daoist discourse.

> In the northern ocean there is a fish named Kun. I don't know how many thousands of *li* is the great size of Kun. It transforms into a bird named Peng. I don't know how many thousands of *li* is the size of its back. In a passion it flies off, and its wings are like clouds hanging from Heaven. When the ocean churns, then this bird heads off to the southern ocean. The southern ocean is the Pond of Heaven.[7]

> The *Qi Xie* [translatable as *Tales of Harmony*, *The Wonders of Qi*, or *Book of All Jokes*] records marvels. In the words of the *Tales*, it says, "When Peng moves off to the southern ocean, the waters are thrashed up to three thousand *li*, it mounts the whirlwind ninety thousand *li* above, and travels for six months before resting."[8]

> (Emperor) Tang asked Ji about this. "The dark ocean of the north where nothing grows is the Lake of Heaven. There is a fish there, whose width is several thousand *li*, and nobody knows how long it is; its name is Kun. There is a bird, whose name is Peng, its back is like Mount Tai, and its wings are like clouds hanging from Heaven. It ascends the ram's horn whirlwind for ninety thousand *li*. When it rises above the clouds and supports the blue Heaven, it heads for the south to arrive at the southern ocean."[9]

The images in these passages attain their full symbolic valences through participation in the binary cohesion of *yin-yang*. There is a symbolic identity between the Peng bird, Fu Xi, and *yang*, and between the Kun fish, Nu Gua, and *yin*. These images call to mind more traditional representations of *yin-yang*, which picture a dragon 龍 (*long*) as *yang* intertwined with a phoenix 鳳 (*feng*) as *yin*: throughout Imperial China, the emperor occupied the "Dragon Throne" and his consort, the "Phoenix Throne." In *Zhuangzi*'s representation, the symbolic valence of *yang* as male bird (Peng) and *yin* as female fish (Kun) changed to *yang* as male dragon and *yin* as female bird, most likely due to the resonance of emperors and dragons. This resonance appears very early in the Chinese tradition: a dragon symbolizes the first hexagram of the *Yijing*, *Qian* 乾, with its six *yang* lines, while a mare symbolizes the second hexagram, *Kun* 坤, with its six *yin* lines. Here, then, the symbolic connections should be understood as *yang*-Peng-dragon/bird, and *yin*-Kun-fish/horse. It is unclear to me how the *yin* codes came to be represented by a phoenix. These images are also related to another traditional representation of two mythical birds, the Feng 鳳 female phoenix and the Huang 凰 male phoenix.

Although the *Zhuangzi* steers clear of explicitly designating the creative qualities of Kun and Peng, they still fully participate in the symbolic attributes commonly found in the origin myths concerning the primordial couple and the sexual generation of the world. Discussing these attributes, Girardot writes, "Where the primary associations for Fu Xi involved the snake-dragon symbolism, the linguistic associations connected to the name Nugua had the meaning of 'snail, frog, water-hole, pond, etc.'"[10] Both Kun and Nu Gua have deep affinities with the depths of a watery, oceanic realm, locked away in darkness from human cognizance, and thus we can have no substantial description of Kun: "I don't know how many thousands of *li* is the great size of Kun." Peng and Fu Xi are celestial creatures soaring above in the skies, fully transparent in the light of day, and we can somehow survey the size of Peng in its celestial appearances and movements: "I don't know how many thousands of *li* is the size of its back."

Throughout these passages, we find several general numerical measurements. In the traditional attributes of *yin-yang*, *yin* is commonly associated with the north, darkness, winter, and water, and *yang* with the south, brightness, summer, and fire. Nu Gua, as we have seen, is associated with snails, frogs, and waters, and the name Kun itself refers to fish roe, ironically having the associations of abundance and generation that come with the hatching of fish eggs, while at the same time referring to something extremely small. Naming something as huge as Kun with a term signifying a miniscule object further works to emphasize the inability of human beings to know and or signify the earliest beginnings.[11] The temporal associations of Peng and Kun are also clear, for it seems that Kun the fish remains unchanged as a fish for six months, then becomes Peng the bird for six months. Instead of the sexual symbolism identified with the primordial couple, we are introduced to the early Daoist notion of transformation 化 (*hua*): the fish transforms itself into the bird, in the same way that *yin* changes into *yang* and *yang* into *yin* throughout the course of the annual progression of the seasons. Presenting the processes associated with natural transformation in these sorts of ways serves as one of the primary means in which the early Daoist writers made thinkable their intimations of the pristine Dao.

These passages from the *Zhuangzi* offer a complex representation of the alternations and transformations of *yin-yang* that work on different levels of signification. These levels include both the very beginnings of the cosmogonic world, and the continued cyclical progressions of time and change in the natural world. The early Daoist writers greatly expand the range of applications of *yin-yang* through conscious redeployments of the ancient Chinese mythological images revolving around snakes, dragons, and birds. However, the symbolic valences of these mythological images are consistently compromised and subtly altered by being put into the service of early Daoist discourse, which primarily employs them only to go beyond what they had traditionally represented. Kun the fish and Peng the bird, metaphorical representations of *yin-yang*, appear for all intents to be the literary creations of the writer of *Zhuangzi* 1, yet their symbolic attributes intimately resonate with those of the primordial couple Fu Xi and Nu Gua. These deviant strategies, relentlessly working to reconfigure traditional symbolic valences through cunning redeployments in accord with their own discursive purposes, represent a hallmark of early Daoist discourse.

The account of Kun and Peng represents neither a cosmogony nor a creation, but it does open a window onto the ways that early Daoist writers can be seen to take up earlier mythological images and motifs for their own ends. It shares with other cosmogonic narratives a depiction of the beginning of the world in terms of a watery chaos that is always already present. Another element they have in common with other cosmogonic narratives is the portrayal of the primordial couple. These two elements—namely, a watery chaos and a primordial couple—represent the primary substantive inheritance from the preexisting mythological fund deployed in the development of early Daoist cosmogonies. Typical non-Daoist tellings of the beginnings, exemplified by the Chu manuscript, begin with descriptions of the already existing watery chaos, and then present the primordial couple. While the manuscript does not depict the origins of the primordial couple, it can be surmised that the existence of the watery world predates them. Further, their activities coincide with the temporal and spatial developments of the world that were carried forward through their sexual activities and their offspring. The primordial couple is not identical with *yin-yang*, rather they use *yin-yang*, in the form of the "hot and cold *qi*," in pursuing their works.

Early Daoist writings, on the other hand, systematically assert an identity between images representing the primordial couple and *yin-yang*, where the two members, whether Fu Xi and Nu Gua, Kun and Peng, the unicorn and phoenix, Heaven and Earth, or man and woman, to name a few instances, all appear as substantive representatives of the larger category of *yin-yang*. With these rhetorical strategies, early Daoist writings also rewrite the generative sequence of the ancient myths. *Yin-yang* comes into being directly from the body of the pristine Dao during the course of the cosmogony, thereby chronologically preceding the existence of the world and all other phenomena. In the early Daoist cosmogonies, *yin-yang* directly give birth to the world by way of their mingling, and this is, unsurprisingly, typically represented in strongly sexual images. All of this had a fundamental effect on later Chinese understandings of the beginning of the cosmos, Daoist and non-Daoist alike. Historically, the early Daoist cosmogonies appear to be the earliest in the recorded writings of early China, and they had deep repercussions on other, non-Daoist literary traditions, as seen for example in the *Yijing* text analyzed in the following section.

THE *XICIZHUAN*: AN ALTERNATIVE COSMOGONY OF THE CONFUCIAN TRADITION

When other traditions started to compose their own cosmogonies, they did not uniformly adopt the sequence established in the early Daoist writings. In brief, the Daoist sequence begins with the Dao; the Dao gives birth to the *qi*, the *qi* gives birth to *yin-yang*, *yin-yang* give birth to the three realms of Heaven, the Human, and Earth, which in turn give birth to the ten thousand things 萬物 (*wanwu*), all phenomena. The Daoist sequence progresses numerically: from One, to Two, to Three, to ten thousand. The *Xicizhuan* 繫辭傳, a Confucian piece produced around the end of the Warring States period and appended to the *Yijing* 易經, presents an alternative cosmogony. Standing as a Confucian cosmogony, it embraces some of the primary elements of Daoist cosmogonies but rejects others. In it, the Dao retains its position as the source from which all things subsequently emerge; in the *Xicizhuan*, however, the most important products of the Dao are not *yin-yang*, but rather the two trigrams, *Qian* 乾 and *Kun* 坤, understood as the pure embodiment of *yin-yang* in the cosmogonic process. The text describes *Qian* and *Kun* in terms that call upon the sexual imagery of *yin-yang*: "The Dao of *Qian* produces the male, the Dao of *Kun* produces the female. *Qian* knows the Great Beginning, *Kun* creates the completed things."[12]

The general cosmogonic view of the *Yijing* is represented by explicit identifications between *yin-yang* and *Qian* and *Kun*: "Now *yin*, now *yang*, that we call the Dao. . . . Their creating images is called *Qian*; their imitating models is called *Kun*."[13] Gerald Swanson gives a clear explanation of this identification in his comments to this passage.

> This passage is alluding to the famous cosmogony passage in Chapter 11 of the *Xicizhuan* and Chapter 42 of the *Daodejing*. The Great Ultimate [the Dao] gives birth to the two principles of *yin-yang*, the two principles of *yin-yang* give birth to the four images. The so-called four images are normally called "the Great *yin*, the Great *yang*, the Small *yin*, and the Small *yang*." Patterns are in the heavens and forms are on the earth. As forms on the earth, they are imitative; as patterns in the heavens, they are creative. Thus *Qian* and *Kun* are merely two aspects of the same thing as *yin-yang* are also two aspects of the same thing: the total undifferentiated unity of all existence. Thus when *yin-yang* alternations create images they are called *Qian*; when they imitate models, they are called *Kun*.[14]

There are three moments of critical importance in the *Xicizhuan*'s understanding of the cosmos. The first establishes the original presence of the Dao as prior to all else. The second occurs when the Dao reveals itself in two modes that manifest in one of several ways: as *Qian* and *Kun*, Heaven and Earth, *yang* and *yin*, noble and base, or man and woman. Each of these manifestations emerges in the form of a pair, in which the first member is always valued more highly than the second. This is because the cosmos of the *Xicizhuan* is structured by an essential and fundamental hierarchical dualism, where the first members of each pair participate in a superiority of value common to each, and the second members participate in an inferiority of value common to each. There is an analogous relation between all first set members of each group, and between all second set members of each group. Probably because of early Daoist notions of an equalizing *qi* that was never absent from their understanding of *yin-yang*, *yin* is systematically named before *yang*; one rarely finds early Daoist writings discussing *yang-yin*. The *Xicizhuan*'s cosmogony, however, marks a radical departure from that Daoist cosmogony, on structural terms, by its tendency to speak of *yang* before *yin* (and this tendency became more and more pronounced in later Chinese writings). This is clearly shown in that text's quotation of the words of Confucius: "The Master said, '*Qian* and *Kun*, do they not constitute the two-leaved gate into the *Changes*? *Qian* is a purely *yang* thing, and *Kun* is a purely *yin* thing.' "[15] Moreover, in the *Xicizhuan*, there is no third term that reveals an ultimate unity between the members as found in the Daoist use of an equalizing *qi* to stand for a fundamental unity of the cosmos. Furthermore, in the *Xicizhuan*, the third moment marks the role of the sages who interpret and understand the hierarchical truths of all existence (revealed in the lines of the *Yijing* text itself), and make those truths known to all people in the construction and maintenance of the good society. Thus, the text adapts and transforms the early Daoist mythological motifs for its own purposes.

The *Xicizhuan* presents an eminently Confucian vision of the beginnings that became standard in the belief systems of the traditional Chinese literati and the imperial ruling houses, particularly from Song times onward. In this view, the cosmos evolves through the production of dualistic hierarchies that model in turn the basic distinction between the cosmic trigrams *Qian*, imaged in terms of the male-father-ruler, and *Kun*, imaged in terms of the female-mother-minister. Images of sexual generation provide a distinctive element of this creative process, which is understandable since this vision remains deeply influenced by the early Daoist versions of the beginnings that were grounded in the sexual intercourse of *yin-yang*. The sexual nature of the reproduction of the different levels of the dualistic hierarchies appears in the opening lines of the *Xicizhuan*: "As Heaven is high and noble and Earth is low and humble, so it is that *Qian* and *Kun* are defined. . . . Heaven creates images, and Earth creates forms; this is how change and transformation define themselves. . . . In consequence of this, as hard and soft stroke each other, the eight trigrams activate each other".[16]

Here, "noble" Heaven and "humble" Earth, both possessing their own creative potencies, represent the world as a whole. Based on the text, it is impossible to know clearly which set of pairs holds the highest value for the *Xicizhuan*'s view of the cosmos: *Qian* and *Kun*, Heaven and Earth, or *yin-yang*. Nonetheless, the spatial and temporal limits of the world are defined by Heaven and Earth, in which the hard and the soft (the unbroken *yang* lines and broken *yin* lines of each hexagram) are produced. The "hard and soft" are then technical terms for *yin-yang* on the microcosmic scale of the text. Holding to the notions of sexual generation, the text states that they stroke 摩 (*mo*) or rub against each other in a kind of sexual embrace, and by this the eight trigrams come to life. The world that is depicted in the *Xicizhuan* is a world structured by the maintenance of a hierarchical structure, where Heaven, *Qian*, *yang*, kings, and husbands have an enduring position of value and superiority over Earth, *Kun*, *yin*, ministers, and wives. The *Xicizhuan* also speaks of the Dao, but here the Dao does not represent the true nature of the world based on a unity of being; rather, the world is founded on and structured by fundamental oppositions, such as that between high and low, which are maintained at every step of the well-ordered world. This vision of the world is very much at odds with the visions put forth by the early Daoist writings, which developed a Dao-based cosmogony grounded on an ultimate, spontaneous unity of being rather than a dualistic, hierarchical structure as seen in the *Xicizhuan*.

ABYSSAL WATERS

Before leaving the analysis of Chinese mythological creatures associated with sexual generation, watery chaos, and the beginnings of the world, I would like to mention a curious text found in the *Guanzi* 管子 collection, entitled *Shui Di* 水地, *Water and Earth*. The title is a misnomer, for the first word, *Shui* (water), was mistakenly written as the word for *Di* (earth), thus giving this piece its erroneous name; throughout it is concerned only with water and not with earth.[17] Allyn Rickett dates it from anywhere between the end of the Warring States and the beginning of the Han. Because it shares strong lines of affinity with sections of the *Huainanzi*, I agree with Rickett's idea that it was likely written at the court of Huainan. Although this is a very late piece of writing in terms of early Chinese chronology, it is noteworthy because it takes up several of the major themes of this chapter and demonstrates deep affinities with the stories of Kun and Peng. Its portrayal of cosmogonic processes is very similar to other, Dao-centered descriptions found in the early Daoist writings. It begins with a discussion of the circulation of water.

> Water is the root of all things and the source of all life. . . . Water is the blood (*xue*) and breath (*qi*) of Earth, and is what flows through the sinews and veins. Therefore it is said, "Water is complete in its substance. . . ." Therefore, there is no place it does not fill and no place it does not reside. It gathers in Heaven and Earth, is stored in the ten thousand things, is produced in metal and stone, and gathers in all forms of life. Therefore it is said, "Water is divine."[18]

The *Shui Di* is not a cosmogony, but parts of it do lend themselves to cosmogonic ruminations. Because it does not discuss the ultimate beginnings of things, there is little mention of the cosmogonic properties of water; it focuses on the generative power of water and its role in the fashioning of life. In three consecutive sections, the text describes the role of water in the sexual coupling of human beings, in the fashioning of two mythological creatures, and in the forming of two peculiar spirits. As for these two spirits, it states that stagnant marsh water give birth to the *qingji*, a creature four inches tall wearing yellow and riding a horse, and that stagnant river water gives birth to the *wei*, a creature having two bodies and one head and looking like a big snake.

In its description of the relation between water, human beings, and the sexual act on one side, and water and the two mythological creatures on the other, the text calls upon the vitally potential creative powers of water for bringing forth life. Things come forth into existence through a process of binary emergence, where one thing comes to exist simultaneously with its opposite or complement. These two beings again unite with each other by water and thereby generate yet more life; these textual descriptions are strongly sexual in diction.[19] The text states: "Humans are water. When the *jing* and the *qi* of male and female unite, water passes between them and assumes form [as a fetus]."[20] The text describes the ten-month gestation period resulting in the birth of an infant and points to many five-phase correlations between the parts of the fetal body.[21] The cosmogonic tenor of the *Shui Di* passage comes forth clearly in the following section.

> Things that hide in dark recesses and are able both to appear and disappear are the ancient tortoise and dragon. The tortoise lives in water but is opened by fire [a reference to turtle-shell divination]. Thus it can tell all things and verify the coming of good times and bad. The dragon lives in water but travels around wearing the five colors. Thus it is a divine spirit. When wishing to be small, it transforms itself into a caterpillar. When wishing to be large, it envelops both Heaven and Earth. When wishing to rise, it reaches the clouds. When wishing to descend, it enters a deep spring. It transforms itself without regard for the day; it rises up or descends without regard for the hour. Thus we call it a divine spirit.[22]

This is another presentation of the primordial couple rendered in images that call upon those used by the Chu silk manuscript describing Fu Xi and Nu Tian and the *Zhuangzi*'s depiction of Kun the fish and Peng the dragon. Although there are no depictions of the time or environment of the very beginnings of things that one expects from a cosmogony, the *Shui Di* continues to share in the complex of early Chinese mythic images and motifs.[23] Another characteristic identifying it as a whole with early Daoist discourse is particularly apparent in the descriptions of the human sex act that use the terms *jing* and *qi* and describe their coming together in the generation of new life. The assimilation of water with *jing* and *qi* first appears in the opening lines of this text previously cited, where water is identified with blood, a common referent for *jing* in the human body, and with *qi*, believed to circulate both in the Earth and the human body. Surprisingly, however, this text never once mentions the Dao and thus never draws a direct relation of identity between the Dao and water. It explicitly develops a description of actively functioning opposites or complementarities—dragon-tortoise, male-female, *qi-jing*—together with a theory of the mechanisms by which they emerge from a prior substance viewed in images of water. The union of these opposites is effectuated through their returning to their source, resulting in the further generation of new life. These mechanisms replicate those that structure the cosmogonic processes as they are consistently portrayed in the early Daoist writings.

In the early Daoist cosmogonies, *yin-yang* emerge from a primordial environment regularly presented in terms of a dark, murky, watery chaos that is sequentially a few steps removed from the pristine presence of the pristine Dao. Once *yin-yang* emerge from this watery environment, they proceed to give birth to the world, with the generation of Heaven and Earth placed squarely after their already established presence. From Heaven and Earth, then, are generated the ten thousand things, including human beings. One of the most celebrated instances of this fully developed cosmology is located in the opening passages of *Huainanzi* 7:

> In ancient times before Heaven and Earth, there was only the Image [the Dao] without body, obscurely, obscurely, darkly, darkly; a vast network of branchings within a desolate silence; misty oceans abyssal and merged, and none can know its gates. Two divines were born from the chaos; their movements fixed the courses of Heaven and laid the lines of Earth. Such emptiness! None know the extent of its limit. Such surgence! None know the subsidence of its movement. These then are the distinctions in the operations of *yin-yang*: their divergence from each other effectuated the settling of the Eight poles, and their separation constituted the formation of the Hard [Heaven] and the Soft [Earth], by which the ten thousand things are brought to embodiment. The diffuse breaths formed animals, the pure breaths formed humans. So, the *jing* (fluid principle) and the *shen* (spirit) are of the provenance of Heaven, and the skeletal bones of the provenance of Earth. The *jing* and the *shen* pass through the gate (of Heaven), and the skeletal bones return to their roots (of Earth).[24]

These celebrated passages offer an unmistakably Daoist cosmogony. In keeping with the ancient images, this passage places the beginnings of the world in a realm of watery chaos where all things have yet to be differentiated, a kind of primal soup. Instead of a primordial couple, there is *yin-yang* to which no anthropomorphic characteristics are attached; the early Daoist writings consistently refrain from positing deity figures.[25] Yet what definitively makes the early Daoist cosmologies significantly different from all of the earlier, water-based mythological presentations is that they place the pristine Dao in a time even before the existence of this watery, abyssal environment. In the early Daoist descriptions, this watery, chaotic and undifferentiated mass itself is the material effect of the cosmic movements of the Dao. For these writers, this watery realm was understood as the manifestation of the breath of the Dao or the wind caused by its motions. *Qi*, a word meaning "breath," "vapor," and "wind," is used to designate this first-generation offspring. It is the cosmogonic "stuff" that gives birth to *yin-yang*. In its associations with breath and vapor, the watery associations of the *qi* become apparent, vapor being the steam of any heated liquid substance. The *qi* itself is the watery mass that defines the absolute limits of what there is.

The importance of water as a cosmogonic symbol used to describe *qi* in representations of the time before time is a striking feature of early Daoist writings. This cosmogonic *qi*, the stuff from which all subsequent phenomena are born, is the physical or energetic manifestation of the pristine Dao as the existent residue born from its movements. The semi-identity of *qi* and Dao allows water to be used to represent both. The Dao itself is commonly imaged in terms of water; representative examples of this identification are abundant.[26] *Laozi* 8 states: "The highest good is like water. Water is good at benefiting the ten thousand things and yet it does not compete with them. It dwells in places the masses of people detest, therefore it is close to the Dao."[27] *Laozi* 4 states, "The Dao is empty, yet when you use it, you never need fill it again. Like an abyss! It seems to be the ancestor of the ten thousand things . . . Submerged! It seems perhaps to exist. We don't know whose child it is; it seems to have preceded Di."[28] This cluster of cosmogonic images (Dao as the original source of all things, *qi* as the life-giving stuff received by all things, and the abyssal waters of chaos) informs these passages from the *Laozi*.

In two closely related passages from the *Zhuangzi* and the *Liezi*, the watery environment preceding the formation of the world is given a more complete depiction. The *Zhuangzi* passage is especially noteworthy not only in that it regards these watery worlds as in some sense cosmologically prior to the formation of Heaven and Earth, but also because it makes clear that these prior realms continue to be accessible to certain human beings. To gain access to these realms allows one to attain a physiological experience of the pristine Dao itself. In the beginning of this passage, the reader is introduced to a shaman named Ji Xian 季咸. He possesses the ability to read the faces of others and tell their future, in particular the date and time of their death. The novice Liezi 列子 takes notice of him and wants him to meet his Daoist teacher, Huzi 壺子, "Gourd Master."[29] Escorted by Liezi, Ji Xian visits Huzi four times, and at each visit Huzi reveals progressively deeper physical identifications of his body with the different phases of the cosmogony, each identification reaching further back in time to the very origins, ultimately culminating in the identification of his body with the pristine Dao itself. Intending to diagnose the fortunes of Huzi by reading these identifications through his face, Ji Xian in the end is reduced to terror and flees.

After the first meeting, Ji Xian tells Liezi that his master has only a few weeks to live. Going in to report to his master, Huzi explains what had happened to Liezi: "Just now, I revealed myself through the patterns of Earth. The sprouts were not violent but were unceasing. That was when he began to see me restricting the impulses of *de*."[30] After the second visit, Ji Xian tells Liezi that his master has recovered. Huzi then says to Liezi, "Just now I revealed myself through the ground of Heaven. Name and substance did not enter, and the impulses came from my heels. That was when he began to see my impulses toward goodness."[31]

After the third visit, Ji Xian says that he can get no reading of Huzi's face, and tells Liezi to ask him to go on a fast. Huzi then explains, "Just now I revealed myself as the ultimate overflow where nothing is ascertained. That was when he began to see me in the leveling of the impulses of *qi*. In the gathering of whirlpools there is an abyss, in the gathering of still waters there is an abyss, in the gathering of flowing water there is an abyss. These abysses have nine names, (I showed) these three."[32] After the final visit, Ji Xian flees in horror from Huzi, who then explains to Liezi, "Just now I revealed myself as not yet having emerged from the ancestor. I wriggled and snaked with him in the Void, not knowing who we were. From this we began to fade away, from this we began to be carried away in the waves. That's why he fled."[33]

Although this passage can be read as demonstrating a confrontation between the arts identified with the shamans and the arts identified with the early Daoists, in which the Daoist side is shown to be victorious, there is something more going on. This passage demonstrates a partial vision of the cosmogony that the *Liezi* passage later fills in. The sequence of the progressively deeper identifications of Huzi himself, consecutively reaching back to ever earlier periods of the cosmogonic process, represents the characteristically Daoist technique for achieving a complete identification with the pristine Dao through physical embodiment. The structure of the passage begins in the realm of the Human, and proceeds backward into the realms of Heaven and Earth. Reflecting the distinctions found in the *Huainanzi* passage cited earlier, Earth is associated with material existence and the bursting forth of the life forces in all beings, and Heaven is associated with spirit and breath, the *qi* that comes from the heels.

Huzi's explanations to Liezi center around the notion of "impulses" (*ji* 機). These impulses refer to the "impulses" of the generative force of life and the world, in other words *qi*. For the realm of Earth, these impulses are directed to the surging forth of vegetative life, while for the realm of Heaven these impulses are directed to goodness, *shan* 善, in the sense of the goodness of existence as such. Huzi further explains that the impulses from Heaven came to him from his heels, and the text here resembles the description of the Genuine Person 真人 (*zhenren*) found in the opening lines of *Zhuangzi* 6: "The breath of the Genuine Person comes from the heels; the breath of the common person comes from the throat."[34] To be able to breathe the breath of the cosmos directly and deeply through specific programs of breath control was a technique of primary importance for achieving the physical unification with the pristine Dao itself. These impulses from Heaven and Earth are the manifestations of the life force identified with the *qi* and seen as the name and substance of generation imaged in the waters of the abysses from which Heaven and Earth were generated. In their spatial and temporal limits, Heaven and Earth create restrictions on this pure potentiality of the *qi* simply by virtue of their brute existence; before their formation, the *qi* enjoyed a complete freedom that was structured only by its rhythm of movement that in time came to form the Nine Abysses. The *qi* envisioned as the surging waters of the Nine Abysses is itself the manifestation of the movements of the pristine Dao.

Going beyond the realms of the Human, Earth, and Heaven, the world is left behind, but this does lead to a realm of nothingness. There is a state of pure potentiality ontologically prior to Heaven and Earth and their shared participation in existential time and locative space, yet continuously open to experience even after they have been established, as evidenced by Huzi's activities. This realm is textually imaged as watery abysses from which all life is generated, standing midway between the pristine Dao and the manifest world. Huzi's embodiment of this realm of potentiality leads to the description of it as the "ultimate overflow 太冲 (*taichong*) where nothing is ascertained." Finally, Huzi achieves a complete embodiment with the pristine Dao in which even the watery abysses have not yet emerged; he is one with the ancestor 宗 (*zong*), which is also described as the Void 虛 (*xu*) and characterizes the pristine Dao of the origins.

The *Zhuangzi* appears to be partial in its descriptions of the Nine Abysses; only three are there presented. The same story is given in a very slightly altered form in the *Huangdi* 黃帝 chapter of the *Liezi*; the only significant difference between the two versions is that the *Liezi* enumerates all Nine Abysses.

> A gathering of whirlpools forms an abyss, a gathering of still waters forms an abyss, a gathering of flowing water forms an abyss, a gathering of flood waters forms an abyss, a gathering of irrigation waters forms an abyss, a gathering of cavern springs forms an abyss, a gathering of returning waters forms an abyss, a gathering of marsh waters forms an abyss, a gathering of collected waters forms an abyss. These are the Nine Abysses.[35]

The addition of the six abysses not named in the *Zhuangzi* serves to fill out the complete description of the cosmogonic environment. The generative potency of these waters is constantly underscored as the vital and material potentiality in the generation of all life, the *qi* as the breath of the cosmic Dao.

These early Daoist writings demonstrate a sustained effort to control discursively the periods preceding the formation of Heaven and Earth. Instead of a murky, watery chaos that roughly describes the beginnings of the world, these writings present specific representations together with designations of, for example, nine watery abysses. These depictions characterize cosmogonic origins in specific ways that have deep ramifications for how human beings are to understand and manage the manifest world. They express a rejection of traditional notions of hierarchy and the current systems of value and structures of social practice. The world envisioned by the early Daoists writers is ultimately resolved in a fundamental unity, rather than in fundamental structures of hierarchical dualisms. The detailed descriptions of the watery abysses play a definite and necessary part in their cosmogonies, which strive to bring out into clear light the relationship between the pristine Dao and the manifest world of human existence. This posited relationship keeps open the possibility of soteriological reversion to a direct identification with that Dao through physical embodiment. This route from here to there leads through the realms of the watery abysses as the pure potentiality standing between the Dao and the world.

There are further important connections between these passages from *Zhuangzi* and those earlier cited from *Laozi* 4.[36] Three important terms are shared by both, namely "overflow" 冲 (*chong*), "abyss" 淵 (*yuan*), and "ancestor" 宗 (*zong*). Both sets of passages envision the Dao in its relation with water, and, at the same time, affirm that through this waterlike quality the Dao is able to give life. The *Laozi* writes that "the Dao overflows" 道冲 (*dao chong*), while the *Zhuangzi* passage names the watery realm as the "Supreme Overflow" 太冲 (*tai chong*). The watery environment in both passages is described as an abyss (*yuan*), and the *Zhuangzi* further cites nine of them. Finally, the *Laozi* says that "the Dao is the ancestor of the ten thousand things," while the *Zhuangzi* passage, describing the unity of Huzi and the Dao, says, "I revealed to him what it was like before I had emerged from my ancestor." Without delving into the issue of textual borrowing, it is nonetheless clear that both are calling upon the self-same body of technical vocabulary and general images specifically identified with early Daoist discourse. Both, furthermore, present alternative, even subversive, perspectives on human origins by discarding commonsense notions of human lineage, seen in their unexpected use of the term "ancestor," that lie at the core of both the Confucian and the ancestral ideology current in early China.

PLACENTAL WATERS

Early Daoist writings commonly rely on images of creatures associated with water to depict water-dominated environments of primordial potentiality and nondifferentiated murk. The antiquity of these images can be traced back to earlier images of dragons, snakes, tortoises, and primordial couples usually associated with themes of sexual generation through the union of opposites or complements. Early Daoist cosmologies assimilate these themes and images to the notion of *yin-yang* in order to represent the ancient symbolic attributes, but their uses of those attributes go beyond the specific mythological valences with which they are identified. The early Daoist writers accomplish this by relying on their own specific notions of the pristine Dao, which itself is often imaged in terms of a similarity with water and the watery environs of the time before time. In turn, these waters are regularly discussed and described in terms of *qi*. One of the primary ways in which the early Daoist writings made their mark on early Chinese society and culture was through the impact of their sophisticated speculations on a cosmogony that preceded gods, dragons, and even water itself. These topics were apparently left largely unexplored by all other discursive traditions of early China. We possess tantalizing pieces of non-early Daoist writing that directly attend to the beginnings of the cosmogony and that see those beginnings as generated directly from the medium of water. These water-cosmogonies strike a transitional phase between, on the one hand, the mythological origins of the world represented by the actions of mythic creatures, dragons and snakes, and, on the other hand, the developed early Daoist visions that go all the way back to the pristine Dao existing as a pure potentiality for everything that is.

The earliest writing that gives a presentation of the very origins of the cosmos and its association with water, together with a presentation of the generation of the world as forming from those cosmogonic waters, is the short piece called the *Taiyi Sheng Shui* 太一生水 (*Taiyi Gives Birth to Water*). It was part of the collection excavated from the Guodian site, dated to the middle of the fourth century BCE. A number of the Guodian writings reveal strong lines of affinity with the then-emerging early Daoist discourse; in fact, the *Taiyi Sheng Shui* itself was bound together with the Guodian *Laozi* in the tomb. Despite this, the *Taiyi Sheng Shui* is not, as such, an early Daoist writing, and many of its central themes stand halfway between water-based cosmogonies and Dao-based cosmogonies.

Until the Guodian excavations, the earliest known writings mentioning Taiyi dated back only to the end of the Warring States, long after the early Daoist discourse had been substantially developed and disseminated. Modern scholars long considered Taiyi to represent only a philosophical notion used by the early Chinese writers to depict the principle of origination. It was thought that Taiyi was elevated to the position of a supreme deity in the period of the Former Han dynasty, when he was named among the recipients of imperial sacrifices. In recent studies, based in large part on the Guodian materials, Li Ling and Don Harper have demonstrated the existence of earlier Warring States religious practices involving Taiyi, and they conclude that Taiyi represented a religious deity already by the beginning of the Warring States. [37] Li writes that "although at present we still cannot know exactly when the worship of Taiyi began—and therefore cannot determine exactly what the relation is with philosophical concepts like Taiyi, Dao, and Taiji—nevertheless, we can still show that in the pre-Qin period Taiyi was already a concept that included the senses of astral body, spirit, and ultimate thing."[38] The *Taiyi Sheng Shui* is now generally recognized as China's first written cosmogony and provides the first indications of a being, principle, or entity predating the existence of the watery chaos. Harper writes, "I believe that *Taiyi Sheng Shui* is best read as a religious cosmogony; that is, the text gives us the oldest Chinese cosmogonic account in which genesis is initiated by a deity."[39] This is the first section.

> Taiyi gave birth to water; water reverted and conjoined with Taiyi, thereby becoming Heaven. Heaven reverted and joined with Taiyi, thereby becoming Earth. Heaven and Earth (returned and conjoined), thereby becoming divinity and luminescence. Divinity and luminescence reverted and conjoined, thereby becoming *yin-yang*. *Yin-yang* reverted and conjoined, thereby becoming the four seasons. The four seasons reverted and conjoined, thereby becoming hot and cold. Hot and cold reverted and conjoined, thereby becoming moisture and aridity. Moisture and aridity reverted and conjoined, stopping with the formation of the year. Therefore, the year is born from moisture and aridity. Moisture and aridity are born from hot and cold. Hot and cold are born from the four seasons. The four seasons

> are born from *yin-yang*. *Yin-yang* are born from divinity and luminescence. Divinity and luminescence are born from Heaven and Earth. Heaven and Earth are born from Taiyi. Thus, Taiyi is stored in water, and circulates through time. [Completing a cycle it begins again, making itself] the mother of the ten thousand things. At one moment full, at one moment empty, it makes itself to be the model for the ten thousand things. This is what Heaven cannot kill, Earth cannot bury, and *yin-yang* cannot become. The gentleman knows that this is called [Dao]. . . .[40]

This passage represents the earliest Chinese textual designation of the original source of the cosmos, Taiyi, preceding even the watery realms from which the cosmogony unfolds; this is the characteristic feature of Dao-based cosmogonies. Interestingly, it simultaneously holds to the earlier water-based models in which Heaven and Earth predate *yin-yang*, demonstrating that the transition from a water-based to a Dao-based cosmogony was still under way. *Taiyi Sheng Shui* portrays the origins of the cosmos from Taiyi, already present, who (or which) gives birth to water. Through a process of sexual generation, water reverted and united with Taiyi, and Heaven was born. Heaven reverted and united with Taiyi, and thus Earth was born. While Heaven and Earth predate *yin-yang*, Heaven is also given at least a chronological priority in relation to Earth. Although nowhere in this piece is Heaven given any kind of moral priority over Earth, it is clear that this sequence could have helped to lay the foundations for the vision that sees the cosmos constructed on inherently hierarchical principles, as expressed for example by the *Xicizhuan* briefly examined earlier; these views were consistently rejected in the early Daoist writings. Following Heaven and Earth, there comes divinity and luminosity, and only then is *yin-yang* born. In contrast, the early Daoist writings uniformly designate the existence of *yin-yang* as coming before the formation of Heaven and Earth, which are often placed together with the formation of the realm of the Human as third member. This sequence gives a more logically sophisticated progression compared to the linear sequence in which one entity gives birth to another one at a time; in the early Daoist sequence, one gives birth to two, and two gives birth to three.

Here, *yin-yang* are not the cosmic force of generation; instead they are simply identified with weather phenomena ("moisture and aridity") and the annual progression of the seasons, a fairly common association for them in numerous non-early Daoist writings. It was only with the development of the early Daoist discourse that *yin-yang* took on a much more vital position in cosmogonies through their assimilating the images of the primordial couple. Nonetheless, the *Taiyi Sheng Shui* presents mechanisms of sexual generation as the central activity of the cosmogonic processes, apparent from the numerous mentions of "joining" 薄 and "to give birth to" 生 used throughout. Although it was probably written more than a century before the *Shui Di*, the generative structure of both writings, grounded on the act of the uniting or conjoining of two opposites or complements through a reversion back to the original source from which both were born, is virtually identical.[41] Another piece of writing that reveals a remarkable similarity to these texts comes in the first section of the *Dayue* 大樂 chapter of the *Lüshi Chunqiu*.

> Taiyi produced the two principles; the two principles produced *yin-yang*. *Yin-yang* changed and transformed, one ascended and the other descended. They united and became models. In the midst of the primordial and undifferentiated chaos, they time and again separated then united, and united then separated: this is called the constancy of Heaven. Heaven and Earth are like a spinning wheel: upon completion they start again, on reaching the limit they again return, and no one is able to completely fathom this. The movements of the sun, moon, stars, and planets are at times very fast, at other times very slow, but the sun and the moon keep their courses separate and thereby stay on line. The four seasons appear in turn, at times hot and at times cold. [The sky] is at times short and at times long, at times [*yin*] is soft and at times [*yang*] is hard. Taiyi commences the production of the ten thousand things, and *yin-yang* transforms them.[42]

The *Taiyi Sheng Shui* reveals striking similarities with the *Laozi*. The maternal qualities ascribed to Taiyi are completely in line with the early Daoist metaphors that describe the Dao, particularly in its mode as mother. *Laozi* 52 writes: "The world had a beginning, which can be considered the mother of the world."[43] Indeed, the lines from the *Taiyi Sheng Shui* describing Taiyi as a mother are virtually identical to the lines of the opening chapter of the *Laozi*. In the following, I give the lines from the *Taiyi Sheng Shui* (T), followed by the relevant lines from the Mawangdui *Laozi* (M), and the received edition of the *Laozi* (R).

T: a. Completing a cycle, [it starts] over again: [we regard this beginning as] the mother of the ten thousand beings.
   b. First it depletes, then it fills: we regard this beginning as the guideline of the ten thousand things.[44]

M: a. That-which-is-not names the beginning of the ten thousand beings.
   b. That-which-is names the mother of the ten thousand things.[45]

R: a. That-which-is-not names the beginning of Heaven and Earth.
   b. That-which-is names the mother of the ten thousand things.[46]

In addition to the remarkable continuity found in the basic sentence structures of these passages, they all also share a concentrated focus on the life-giving qualities of generation by birth ascribed to the source of all things, whether this is called Taiyi or Dao. These lines from the *Taiyi Sheng Shui*, indeed the piece itself as a whole, have a tremendous influence on the formation of early Daoist discourse, and it marks a dramatic step forward in the consolidation of that discourse.

The myths concerning Fu Xi and his female counterpart describe the formation of the world from the earliest periods. These two deities made use of the substances available to them in the physical world, as well as from their own sexual activity, to establish the various things of the world, yet it is very hard to consider these figures as creator deities *tout court*. Although Taiyi may be considered in many important ways as a creator deity, a view that archaeology might help to establish in the future, the early Daoist writings never take that option; deity figures are virtually nonexistent in them. A great point of difference from the Taiyi depicted in the non-Daoist archaeological and textual records and the Taiyi represented in the Guodian text is that if the latter is a deity, then she is a female deity; the *Taiyi Sheng Shui* is very clear about calling it a mother. Although the early Daoist writings could be considered subversive in their giving a seeming priority to the mother rather than the father, images of the Dao as a mother and the generation of life by birth are a fundamental characteristic of all Daoist discourse, both early and late.

Metaphors and images of water used to represent the nurturing functions of the Pristine Dao allowed further possibilities for relating the Dao to all things by employing the watery, fluid associations of the processes of birth and gestation. Early Daoist writings explicitly adopt the images and metaphors of birth instead of creation ex nihilo. The world and all life were not created from nothing but from the body of the Dao, and the Dao continues to remain present with and in all things. Mawangdui *Laozi* 51 writes, "Dao gives birth to them, nourishes them, matures them, completes them, rests them, rears them, supports them, and protects them. It gives birth to them but doesn't make them dependent; it matures them but doesn't rule them."[47] Creator-figures are appropriate to myths of creation, but the world envisioned in the early Daoist writings is born, not created, and nothing is attributed to creators. Further, creator-figures exercising their will allow for the ascription of moral intentions to the world that is the responsibility of human beings to maintain. Early Daoist writings, however, consistently and rigorously reject all moral intention lying behind the cosmos; this is clear from such passages as *Laozi* 5 that deny the cosmic value of virtues such as *ren* 仁, humaneness: "Heaven and Earth are not *ren*, they regard the ten thousand things as straw dogs."[48]

Taiyi sometimes is also used by early Daoist writers as one alternative designation for the pristine Dao; in fact, one of the most common appellations for this Dao is the single character *yi* 一, One. Current debates concerning whether or not Taiyi in pre-Han times is a deity figure in the full sense of the word notwithstanding, the sense of Taiyi given in the *Taiyi Sheng Shui* does not closely resemble that of a common mythological creator, and therefore I am inclined to treat Taiyi as already an early metaphor for the Dao itself. This referential relation between Yi 一, Taiyi 太一, Taiyi 泰壹, Dayi 大一, Da 大, and Dao is seen in various writings of early China, including the *Lüshi Chunqiu*, the

Guodian *Laozi*, and the Mawangdui *Laozi*. Indeed, the *Taiyi Sheng Shui* does not literally write Taiyi, but rather Dayi; Li points out that "大一, 泰一, and 泰壹 are variant forms of Taiyi's name."[49] A clear example of these linguistic relations between Dao and Da comes from *Laozi* 25. This passage is remarkable not only for showing this referential relation, but also because it presents this relation by calling upon the metaphor of the Mother in describing the life-generating force of the pristine Dao, which allows us to draw a further textual association between the two writings.

> There was something formed out of chaos, born before Heaven and Earth.
> Quiet and still! Pure and deep! It stands on its own and doesn't change.
> It can be regarded as the Mother of Heaven and Earth.
> I do not yet know its name; I style it Dao.
> Were I forced to give it a name I would call it Da.[50]

Laozi 14 states:

> We look at it but do not see it: we name it the minute.
> We listen to it but do not hear it: we name it the rarefied.
> We touch it but do not hold it: we name this the level and smooth.
> These three cannot be examined to the limit, thus they merge together as one.
> 'One'—there is nothing bigger above it, and nothing smaller below it.
> Boundless, formless! It cannot be named, and returns to the state of no-thing.[51]

These several passages from the *Laozi* should put to rest arguments claiming that Taiyi and the Dao occupy different referential positions in the separate early Chinese traditions, that Taiyi as a deity involved with sacrifice and other religious practices belongs to a category completely separate from the pristine Dao as represented in early Daoist discourse. I am not claiming a strict relation of identity between the early Chinese religious deity named Taiyi or Dayi, the slightly later philosophical principle named Taiyi, and the pristine Dao of the early Daoist writers. There is, nonetheless, enough leeway in their fields of signification capitalized on by the early Daoist writers such that they were able to take advantage of a certain assonance among these designations that aided the development of their cosmogonic speculations. In other words, I do not read the *Taiyi Sheng Shui* as representative of early Chinese cultic practice or as simply representative of early Chinese cosmological speculation. I read it rather as one of the earliest representations of an emerging early Daoist discourse on the cosmogony that presents a complete vision of the origins of the cosmos.

Early Daoist cosmogonies are constructed through images of the pristine Dao existing in isolation before the very origin of things; metaphors of the Dao describing it in terms of the One, the Void, and the Mother; portrayals of the cosmogonic beginnings in terms of water symbolism from which all things take form; and depictions of the cosmogonic processes in terms of *yin-yang*. Many writings from early China declare themselves as Daoist simply by using these cosmogonic images and metaphors. The following two passages are instances of the ways in which the early Daoist writers worked each of these elements into their general portrayals of the cosmogony. The first comes from the first sections of the *Daoyuan* 道原 piece from the Mawangdui corpus, and the second comes from the first sections of the *Yuandao* 原道 chapter of the *Huainanzi*.

At the beginning of constant nonexistence,
Totally the same as the Great Void,
Vacuous and the same, it was the One.
Being the One constantly, it was nothing more.
Misty and blurred,
It did not yet possess light and dark . . .
It filled up all within the Four Seas,
And embraced what was outside them.
In *yin* it was not rotted,
In *yang* it was not scorched.
It took One as its measure and did not change . . .
The One was its appellation,
The Void was its dwelling.[52]

The *Yuandao* says:

The Dao shelters Heaven and supports Earth,
Extends beyond the four directions and opens up the eight points of the compass.
It is high beyond reach and deep beyond reckoning.
It envelops Heaven and Earth and gives to the yet formless.
Flowing from its source it becomes a gushing spring:
What was empty slowly becomes full;
From the watery chaos it surges forward:
What was murky slowly becomes clear. . . .
[The Dao] stretches over the four cables holding up the heavens, and harbors *yin-yang* within.
It broadens the cosmos and brightens the sun, the moon, and the stars.[53]

The *Yuandao* continues its discussion with a sustained homage in honor of water, which is quoted in part in the following.

> Of all things in the world, nothing is softer and weaker than water, yet it is so great that it cannot be reckoned, it is so deep that it cannot be fathomed. Its length extends into the limitless, its distance submerges into the boundless. Waxing and waning, expanding and decreasing, it merges with the incalculable. Ascending to Heaven it becomes rain and dew; descending to Earth it becomes moisture and dampness. If the ten thousand things do not get it they will not be born, if the hundred affairs do not get it they will not be complete. . . . Thus, while being neither private nor public, it inundates with a thunderous roar, expansively coursing through Heaven and Earth. While having neither left nor right, it circulates in all directions, and it finishes and starts with the ten thousand things. This is called its supreme *de*. The formless is the great progenitor of things; the toneless is the great ancestor of sound. Its son is light, its grandson is water: both are born from the formless. Light can be seen but not grasped, water can be made to comply but not destroyed. Thus, among things that have an image, none is more honored than water.[54]

Cosmogonic environments in the early Daoist writings are described via metaphors and images of water. The opening passages from the *Jingshen* chapter of the *Huainanzi*, discussed in an earlier section of this chapter, depicts this watery environment in terms of the crashing, oceanic waters of chaos; the story of Huzi from *Zhuangzi* 7, also discussed in an earlier section, depicts this environment in terms of the Nine Abysses. Yet another common way that early Daoist writings portray these waters is in terms of biological metaphors of the womb and the placental sac that gives birth when it bursts open.[55] The image of water when applied to the pristine Dao is regularly depicted as an embryonic environment where fluidlike vitalities undergo the process of gestation. The early Daoist writings commonly liken the Dao to a Mother; each new stage of the cosmology is to be understood as the offspring of the preceding in a family lineage that goes back all to the way to the pristine Dao. Thus, the Dao is often referred to as the Mother (and also, although rarely, as the Father) as well as the ancestor of all things. The family metaphor is of prime importance for understanding the role of human beings and the soteriological possibilities involved with locating and identifying with the Dao present within from birth. *Laozi* 21 states:

> As for the nature of the Dao—it is shapeless and formless.
> Formless! Shapeless! Inside there are images.
> Shapeless! Formless! Inside there are things.
> Hidden! Obscure! Inside there are vitalities.[56]

These images of the Dao as a placental sac deeply resonate with the story of Hundun that immediately follows the story of Huzi in *Zhuangzi* 7. Note here the portrayal of the emperors of the South and North that immediately call to mind the role of *yin-yang*.

> The Emperor of the South Sea was Shu [Impetuous], the Emperor of the North Sea was Hu [Abrupt], and the Emperor of the Center was Hundun. Shu and Hu periodically met in the land of Hundun, and he treated them with great kindness. Shu and Hu were discussing how to repay Hundun's virtue: "All men have seven holes through which they look, listen, eat, and breathe; he alone doesn't have any. Let's try boring them." Each day they bored one hole, and on the seventh day Hundun died.[57]

This story gives a playful characterization of the cosmogonic beginnings when everything that existed was present within the single body of the Dao, exactly like the placental sac and gourd. The coming-to-be of the world occurs in this story as a rupture and deterioration of the initial peace before things began to separate. These images would play an important role in later Daoist thought and practice in which sacs and gourds were assimilated to the life-giving Dao; the notion was extended to alchemical stoves and cinnabar fields in which elixirs were made, and to the gourds that wandering Daoists carried over their backs and into which they crawled at night. All these images represent the pristine Dao before the rupture and symbolize the primordial source to which the Daoist adept desires to return in uniting with the Dao.

These early Daoist cosmogonies laid the foundations for the particularly Daoist worldview and understanding of reality. The transition from the earlier water-based cosmogonies to the Dao-based ones was mediated by the images of Taiyi exemplified in the *Taiyi Sheng Shui* and the *Dayue* chapter of the *Lüshi Chunqiu*. They helped to establish the view of a more or less continuous generation of the many stages of the cosmology, from the coming-to-be of the world culminating in the existence of all living phenomena. The early Daoist writings uniformly embrace birth as the primary cosmogonic image, in which all that comes into being is seen as being born from the internal being of something preceding it, and the further generation of new life comes about by a reversion back to the source from which it was born. Everything necessary for the birth of each new thing is already available in raw, potential, and genetic form in the body of that from which each thing comes forth. However, the pristine Dao, like the Mother's womb, already has the totality of necessary ingredients needed for the generation of all successive levels and beings, albeit in nondifferentiated form within its own cosmic body.

Early Daoist cosmology represents the processes of the continuous refinement of the elements making up all things; nothing is ever added to the world that is not already present in germinal form. Further, everything reverts back to the Dao upon being used up, and thus nothing is ever lost. All of these qualities of the life-giving functions of the Dao are commonly referred to as being like the life-giving functions of water whose attributes and spiritual qualities were continuously extolled in the early Daoist writings. *Laozi* 78 is yet another instance of this high regard for water, combining images of the life-giving power of water with its powers of purification in connection with the Sage. The power of the Sage is like the power of water: because of their purity, they are able to wash away the impurities of the world.

> In the whole world, nothing is softer and weaker than water. And yet for attacking the hard and strong, nothing can beat it, because there is nothing you can use to replace it. That water can defeat the unyielding, and that the weak can defeat the strong: no one in the world doesn't know this, and yet no one can put this into practice. For this reason, the words of the Sage say, "I take on myself the disgrace of the state: this is being Lord of the Altars of Earth and Grain. I assume responsibility for all ill-omened events in the state: this is being King of the World."[58]

Chapter Three

Early Daoism and Cosmology

THE HARMONIOUS WORLD

Early Daoist cosmology is structured by a specific vision of the world as an organic complex of all that exists. Its most common technical designation for this complex is the three-part notion of *tian-di-ren* 天地人, literally Heaven-Earth-Human. I refer to these as the realm of Heaven, the realm of Earth, and the realm of the Human. These three do not designate self-subsistent and impending locations within the cosmos as a whole, but rather parallel and interpenetrating realms characterized by different modes of being. The complex totality consisting in all that is signified by Heaven-Earth-the Human constitutes the world as such. More regularly to be found in the writings are references to the natural, environmental world either as Heaven and Earth 天地 (*tiandi*), or simply Heaven 天 (*tian*). The natural, environmental world is also at times referred to as Nature 自然 (*ziran*), or that which is spontaneously so; references to the sociopolitical and cultural world occur in the term Under Heaven 天下 (*tianxia*), or simply the Empire. In all early Daoist cosmology, there is a clear continuum from the absolute beginnings, when there was nothing but the pristine Dao existing in a state of pure potentiality prior to the existence of the world, to the individual existences of phenomenal things including humans. In these writings, all designations for the world locate it, directly or indirectly, in the latter stages of the sequential unfolding of the stages of the cosmology. Early Daoist cosmology is intimately related to soteriology because they are virtually two sides of the same coin, and it is particularly difficult to separate them. It is only after the world has come into being, at the moments when the three realms have attained some measure of cosmological completion, that the soteriology kicks in to complete the world in an ultimate way.

The world comes to be from within the unfolding of the cosmological sequences; the entire process is recognized as a progressive movement from wholeness to fractured separations, in which the soteriological goal is envisioned as a return to the original wholeness, necessitating a reverse progress backward through the sequences. Other and earlier writings from early China present general frameworks of the cosmos consisting of different levels and open to human passage; in them, shamans 巫 (*wu*) are shown traversing these levels at will. They possess distinct modes of access to separate realms: vertical ones involving journeys of ascension, and horizontal ones involving journeys to the edges of the world.[1] These non-early Daoist writings, however, do not give evidence of a developed, technical vocabulary used to define the cosmologically spatial and temporal qualities of the world; it is not until the appearance of the *Zhuangzi* that we begin to see a rigid technical vocabulary being employed in this way. *Zhuangzi* 23 states:

> There is substantiality, and yet it has nowhere to reside.
> There is duration, and yet it has no beginning or end.
> That from which something can emerge but has no apertures is substantial.
> What is substantial but has nowhere to reside is space.
> What has duration but has no beginning or end is time.[2]

These are very developed notions of time and space, and are characteristic of the kinds of concerns that inform early Daoist cosmological writings; one is hard-pressed to find these kinds of concerns regularly discussed in non-Daoist writings from early China.

The formation of the realms of Heaven, Earth, and the Human mark one stage in the genesis and ongoing existence of the cosmos as a whole. The world then represents one stage of the more widely encompassing activity of the cosmos understood as the processes of the Dao. The Dao itself consists of several related modes of being, but there are two primary ways for envisioning them. The first is in the modes of that-which-is-not 無 (*wu*) and that-which-is 有 (*you*); the second is in the modes of seven parallel and interpenetrating realms or states in which the Dao cosmologically manifests itself, namely *qi*, *yin-yang*, Heaven-the Human-Earth, and, finally, the ten thousand things. Early Daoist soteriology, on the other hand, thus can be said to begin with what is at hand—namely, the individual human being in the world—and, through a progressive series of reversions, to work backwards through the realms constituted by the Human, Earth, Heaven, *yin-yang*, and *qi*, finally to culminate in the complete identification with the pristine Dao of the beginnings.

One of the more intriguing aspects of the early Daoist vision of the world is that it involves something far different from the linear model implied by the previous description. The cosmological formation of the world is an ongoing process with multiple beginnings and multiple completions; spontaneously begun time and again, the process reaches completion only through the third term of the triad, the Human. In other words, the world is brought to completion only by the correct and effective participation of the realm of the Human, but this attained completion is continuously maintained solely through the constant renewal of the harmony of the world, rejuvenated by its intimate unity with the full power of the pristine Dao. The correct and effective participation of the realm of the Human is specifically depicted through powerful representations of the figure of the Sage 聖人 (*shengren*) who successfully completes the soteriological return to the pristine Dao, thereby opening the way for its full presence to radiate directly in the world.

WAS THERE AN EARLY DAOIST COSMOLOGY BEFORE THE *LAOZI*?

Cosmology and soteriology are the two foundational pillars upholding the early Daoist visions of the body and the world. The fundamental concerns for these writers issue first and foremost from the question of the relevance of the world for humans in their effort to embody the Dao. However, it is not just any human being who embodies the Dao, but specifically the Sage; nonetheless, every human being fully possesses the potential for becoming a Sage. The transformation of a human being into a Sage initially commences through that process whereby any human being in general comes to achieve a realization of that aspect of the pristine Dao that is spontaneously and naturally inherent in the physical body from the very moment of the conception of the embryo. The way that the human being, at first simply present as the fetal embryo, and the pristine Dao come to inhere is precisely by way of the world itself. In other words, the world mediates the relation between the human body and the pristine Dao, and at no point can the world be thought to be absent from this relationship.

When human beings are born, they enjoy a complete physical harmony with the Dao, but over time and due in part to the processes of socialization, this harmony gets displaced. The possibility of returning to a state of supreme identification with the Dao is the subject matter for that part of early Daoist discourse that I identify as soteriology; this soteriological vision is itself made possible by the ways in which early Daoist discourse depicts the actual structure of the world and the working relationship among all of its component parts. This is presented in those parts of the discourse that I identify as cosmology. The question of the world, including the issues of its formation, subsistence, and soteriological potential, cannot be resolved into a transcendentalist view whereby the world stands between human beings and the pristine Dao: human beings are not a kind of projectile attempting to shoot clear of their ontological reality in the effort to identify with an absolute principle standing somewhere outside and beyond.[3] The world as such is not amenable to these sorts of bracketings or Husserlian epochés, because it represents the most vital field in which human beings can encounter the full presence of the pristine Dao.[4] The world stands in the direct line of development from the absolute cosmogonic beginnings to the ontological here and now of the present moment, and within the terms of early Daoist discourse it is inconceivable that it could be transcended, bracketed, or negated. In order to understand the soteriological potential of the world and the position of human beings therein, it is first of all necessary to contextualize early Chinese cosmology more generally and its relation to the emergence of early Daoist discourse.

The second section of the *Taiyi Sheng Shui* is one of the earliest writings to depict the presence of the world in conjunction with the pristine Dao.[5] At the same time, it presents an early understanding of the soteriological possibilities for humans in the world. In it we find what might be called a pre-early Daoist depiction of the relations among the Dao, Heaven and Earth, and the exemplary human, the Sage.

> The soil below is what is called Earth. The air above is what is called Heaven. Dao also has its appellation. May I inquire its name? He who intends by way of the Dao to take up the project must rely on its name, and that is why the project will be completed and the body will last. The Sage also relies on its name to take up the project, and that is why the merit will be completed and his body will not be harmed. The names of Heaven and Earth are established together. . . . The Dao of Heaven honors weakness by cutting down completions to benefit what grows. The strong is attacked; the [ ] is punished. . . . [Heaven is somewhat lacking in] the northwest, what lies underneath it is high and strong. Earth is somewhat lacking in the southeast, what lies above it is low and soft. What is lacking above becomes excessive below. What is lacking below becomes excessive above.[6]

Although this section of the *Taiyi Sheng Shui* presents a precise picture of the cosmological structures of the world, nowhere is the triad of Heaven-the Human-Earth directly mentioned. In it, however, are found the earliest textual foundations for the establishment of that three-part view of the world. According to the *Taiyi Sheng Shui*, outside of the Dao there are three different components at work in the formation of the world: Heaven, Earth, and the Sage.[7] Coupling the Sage to Heaven and Earth creates for the writer a limiting constraint over the contours of what this cosmology can do, and this may in part explain why the formation of the world is not here depicted, for how can the Sage stand at the beginning of things? On the contrary, the text depicts the completion of the world, a much more central concern to the writer. Recalling for the moment my discussion of the first section of the *Taiyi Sheng Shui* in the previous chapter, the two sections can be read as standing in a relationship of complementarity: the first section depicts the cosmogonic beginnings, while the second section depicts the cosmological completion. The completion of the cosmology is what is referred to in the phrase "taking up the project" 從事 (*congshi*) and lies under the purview of the Sage. The task of completing the world lies entirely with the Sage, who alone possesses the necessary prerequisites for the job.

The world consists of a triad of Heaven, Earth, and the Sage; here the Sage represents the exemplary human being and the necessary and final component required for the world's completion. The major alteration to this model, once it came into the hands of the early Daoist writers, was to replace the Sage, representing the exemplary human, with the Human as the third member of the triad making up the world that nonetheless still necessitates the actions of the Sage in order to be brought to completion. But this alteration did not radically alter the basic framework of the cosmological model set forth in the *Taiyi Sheng Shui*. The substitution of the Human for the Sage is readily understandable when one realizes that the early Daoist notions of the world include more than the natural world of Heaven and Earth; the model also needed to include the cultural world of the realm of the Human, since it is a crucial component of the entire project brought to completion by the Sage. At the same time, one could make a reasonable argument that another factor in the substitution of the Human for the Sage, or at least an important consequence of it, is that it makes the vision of the world cohere, at least theoretically, with the early Daoist belief that the potential for sagehood is inherent in every human in that it does not specify any single individual in particular.

The first section of the *Taiyi Sheng Shui* already has established the cosmogonic role of the Dao through the explicit identification of it with Taiyi. Underscoring the distinct transition from cosmogony to cosmology, the second section abruptly changes its word usage from Taiyi to Dao in speaking about the cosmic source of all that exists. Although humans exist in the world, and the world reliably subsists from moment to moment, at least according to the commonsense view of things, still it is only a work in progress that perpetually awaits its own completion. The completion of the world is maintained as a soteriological possibility, and the Sage realizes this possibility in full measure. However, this possibility is contingent on the intimate relationship shared by the Dao and human beings, with the realms of Heaven and Earth standing, in a sense, as the conduit or passage for the pristine Dao to the realm of the Human; if the Dao would cease to be present in the world, all life together with the world would come to an end. Humans, although born with the full presence of the Dao at the start of life, easily lose the Dao through socialization and other processes of displacement. If it were not for the Sages, those individuals who consciously cultivate the Dao for themselves and the world and ensure that the Dao will continue to remain present in the world, the realm of the Human would close off, swallow up, and exhaust the creative potential of Heaven and Earth that allow for the life-giving presence of the Dao.

The relation between the Dao and the world is twofold. First, everything that exists, including the world, emerges from the Dao as the cosmogonic matrix of generation. The first section of the *Taiyi Sheng Shui* consistently names this generation as "birth" 生 (*sheng*) in depicting how all things emerge from the Dao. Second, the Dao not only gives birth to all things, it also remains continuously present throughout the processes of world-formation and world-completion.

Human beings stand in the latter stages of a direct genealogical relation with the pristine Dao. At birth, the infant body fully possesses the Dao and can continue to embody it throughout the duration of life, but almost invariably that intimate contact gets displaced with the onset of socialization. That intimate unity can be once again possessed, and the way to it lies within the world itself; before that can happen, however, a process of interiorization is first of all necessary to undo the damage caused by the displacements of socialization. This process of interiorization is consistently represented in the early Daoist writings as a movement of reversal or a return to the beginnings, and this soteriological move of return can be characterized as the process of tracing the Dao, what the *Taiyi Sheng Shui* calls "to rely on its name" 託其名 (*tuo qi ming*). *Laozi* 14 offers an example of this: "Hold onto the Dao of antiquity in order to manage that-which-is in the present; the ability to know the beginnings of antiquity is what is called the genealogy of the Dao."[8] These images manifestly preclude any metaphysical presuppositions that would describe this return movement back to the Dao via transcendental acts that would bracket the world or seek to surpass or negate it. These are the ideas informing early Daoist cosmology, and in large part explain the heavy concentration in these writings given to formulating the relation of the world to the pristine Dao.

The closing lines of the *Taiyi Sheng Shui* discuss the material structure of the world and some of the manifest changes it underwent after its first formation. In early Chinese myth, one finds stories about four pillars holding apart the sky from the land.[9] These stories sometimes describe a great battle between Zhuan Xu, a great emperor of the distant past, and Gong Gong, a fierce adversary who attempted to usurp his throne. In the course of this battle, Gong Gong butted against the northwest pillar, causing one corner of the Earth to tilt up, and the corresponding corner of Heaven to sag down. This story in part explains the ecliptic of the night sky as viewed from the ground, and by presenting these views as cosmology rather than mythology, the *Taiyi Sheng Shui* formulates the manifest and material structure of the world in general accordance with what is conceivably a wider early Chinese empirical view of reality. These sections of the *Taiyi Sheng Shui* indicate the cosmological possibility of returning to the beginnings in order to trace, locate, and identify with the Dao in all things by "relying on its name." The phrase does not imply that the absolute nature of the Dao can be known once and for all by way of names and language, but only that its presence can be grasped by acts of provisional naming, thereby initiating the entire movement of return.

These ideas are in seamless accord with the *Laozi*'s discussions that expound on naming and knowing the Dao; *Laozi* 21 states: "From antiquity to the present, its name has never left."[10] *Laozi* 25 states: "I don't know its name, but I call it 'the Dao.' I provisionally name it 'the Great.'"[11] *Zhuangzi*

25 also offers a highly illuminative passage about the limit or adequacy of naming the Dao. Here, Taigong Zhou 太公周 (Grand Duke Harmony) explains that all things come from a single, undivided origin in the pristine Dao, but the way that humans use names causes that original harmony to break apart. Words can only be used to designate parts, but no words can be used to name the unity. Xiao Zhi 小知 (Little Knowledge) asks whether it is permissible to use the term "Dao" to refer to all that is; Taigong Zhou says "No" and continues his response.

> Now, the number for counting things does not stop at ten thousand, yet what is spoken of as "Ten Thousand Things" is used as a provisional name for reading the great number. Like this, "Heaven and Earth" are used to refer to the greatest of shapes, and "*yin-yang*" is used to refer to the greatest of breaths. "Dao" is used to refer to them both in general. . . . The Dao cannot be said to be "that-which-is," nor can it be said to be "that-which-is-not." What the Dao uses for a name is provisional for the purpose of action or function. The Dao is the limit of all things, and words and silence are inadequate to convey this.[12]

Zhuangzi 25 asserts a nominalist or functionalist designation of the Dao, and this view of naming accords well with the general nominalist views of both the *Taiyi Sheng Shui* and the *Laozi*. According to each of these writings, one relies on (*tuo* 託 or *qiang* 強) operative terms, such as round-off figures or provisional names, in order to gain either a nominalist or functionalist grasp of the Dao and its projects, thereby allowing the Sage to initiate the cosmological completion.

The *Taiyi Sheng Shui* provides an initial cosmological structure for envisioning the world. At the heart of this cosmological structure is a specific soteriology constructed from the activities of the Sage as the one who has been able to rely on the Dao. This writing demonstrates the possibility for the Sage to complete the project. Although the exact nature of this project is not explicitly stated, early Daoist writings, and particularly the *Laozi*, construe this project as the achievement of a second-order completion of the world by way of effectuating a renewed union between the realms of Heaven, the Human, and Earth. In addition to and resulting from this, the *Taiyi Sheng Shui* declares that there are radical consequences directly bearing on the physical body of the Sage; it says that "the body will endure" 身長 (*shen chang*) and "the body will not be damaged" 身不傷 (*shen bu shang*). These are very early indications of the central position that the early Daoist writings consistently give to the fundamentally physical nature of getting, embodying, or unifying with the Dao, and they provide the basic ideas underlying the later practices described under the rubrics of inner cultivation and nourishing the body. The *Taiyi Sheng Shui* leaves these two tantalizing possibilities, namely the completion of the project and the enduring body, dangling without further pursuing the soteriological consequences opened up by these ideas. The early Daoist writings embraced the chance to explore and exploit these ideas by bringing a sustained focus to bear on the relation of the body to Heaven, Earth, and the Dao. In the next section, we will see that the two core issues of the *Taiyi Sheng Shui*—namely, a cosmology built around the pristine Dao and a soteriology built around the Sage—would go on to play a primary role in early Daoist discourse, as is evidenced already in the *Laozi*.

THE HIDDEN SAGE IS NOT A PUBLIC KING

Although it is possible that the origins of early Daoist cosmology are directly attributable to the ideas put forth in the *Taiyi Sheng Shui*, I find that it makes more sense to see those origins gradually taking shape from a general mythological sensibility that is indicated in discrete and independent writings such as the *Taiyi Sheng Shui*. The early Daoist writings adopted and transformed the kinds of ideas found in it, including the idea that the world formed directly from the pristine Dao, and the idea that the world can be represented in terms of Heaven and Earth. Heaven and Earth can represent the world only in its initial stages of emergence; the completion of the world necessitates a third term or agent, the Sage.

The *Laozi* partially adopts this model, but also transforms it in some important ways. First, it substitutes the Human for that of the Sage, thereby demonstrating its conception of the world in terms of three realms of being: Heaven above, Earth below, and the Human in between. This substitution did not in any way alter the project of the Sage, which is still a necessary requirement for the completion of the world. Second, the *Laozi* for the first time discusses the position of the King in relation to the Sage, but it consistently subordinates the King to the Sage. The King appears virtually irrelevant to the Sage's cosmological project of completing the world; in the *Taiyi Sheng Shui*, which had already provided the basic structure of this project taken up by the *Laozi*, the King was not even mentioned. The *Laozi* and other early Daoist writings, while not hesitating to alter the model of the *Taiyi Sheng Shui* in important ways, never tampered with the basic mechanism whereby the Sage completes the project of the world. This is important to keep in mind, because only by doing so can the *Laozi*'s religious vision be properly situated in relation to its political philosophy.

The *Taiyi Sheng Shui* discusses the role of the Sage in bringing the world to cosmological completion, and states that this has definite advantages for the body of the Sage: "The Sage in taking up the project also relies on this name, and when this occurs the project will be completed and the body will not be damaged."[13] The "project" refers to the cosmological completion of the world, and these ideas continue to maintain a central position in the *Laozi*. In the *Laozi*, the term "completion" 成 (*cheng*) is often associated with the "project" and adds an even stronger emphasis on the Sage's role in the cosmological completion of the world, but the term takes on bigger dimensions in other passages. *Laozi* 47 states, "Thus the Sage knows without going, names without seeing, and completes everything without doing."[14] *Laozi* 63 states, "Thus the Sage's ability to complete the great is due to his not acting great, therefore he is able to complete the great."[15] This passage is also repeated virtually verbatim in *Laozi* 34, but it names the Dao as completing the great. As will become clear, the phrase "to complete the great" means to complete the harmony of the three realms of the world. *Laozi* 17 states, "When the merit is completed [by the Sage] and the project is fulfilled, the common people all will say, 'We are so naturally.'"[16] In this passage, the Sage's completion refers to "merit" 功 (*gong*) rather than to the world, but this demonstrates a further sophistication of the previous ideas, not a departure from them. In early Confucian discourse, this term is seldom used. After the imperial adoption of Confucianism in the Former Han, the term takes on more and more the meaning of service well rendered to the emperor or court. In Daoist usage, "merit" carries the striking contrast of cosmological construction. Later Daoist texts will either internalize this notion of merit by identifying it with the placement of deities within the body of the adept, or ritualize it by identifying merit with the construction of the sacred altar.[17]

Early Daoist writings generally distinguish the Sage and the King, although there are instances where the King becomes a Sage; there are no instances that I can find where a Sage becomes a King. The Sage and the King were commonly identified in the non-Daoist writings of early China, most notably in the Confucian, Mohist, and Legalist writings, but the early Daoist writings never shared this view. Because in them the Sage and the King are sometimes associated without being identified with each other, many modern scholars mistakenly have identified the soteriological representative of the realm of the Human with the King. To some degree, this is because later Confucian writings often tried to appropriate the *Laozi*'s insights for their own political ends, and for those writers the King indeed was the Sage. These Confucian readings of the *Laozi* heavily influence many modern interpretations of the *Laozi* that emphasize the political reading that gives priority to the King. In the great majority of introductions found in modern translations, one generally reads that the *Laozi* is a text of governing strategy, with the King representing the ideal reader, and they say it is about the king, too. A very typical version of these ideas is found in the introduction to a recently published collection of essays on the *Zhuangzi*:

> The *Daodejing* is primarily a *political* treatise. As such, it is the portrait of the ruler who, emulating the regularity of nature, sets broad political and social conditions for the pursuit of personal realization. The sage-ruler is thus referred to as the "model of the world" under whose organization the people are at leisure to pursue their own realization.[18]

This is most certainly not the case, because although the King is commonly mentioned in the *Laozi*, true authority is attributed to the Sage, who alone embodies the effective ability to transform the world. The Sage is the genuine source of order and harmony, not the individual holding political position at the top of the political hierarchy. The traditional readings often cause modern readers to misrecognize this crucial aspect of the *Laozi* and early Daoist writings more generally. Modern scholars one-sidedly claim that the *Laozi* proffers a distinctly political philosophy, but in doing so they overlook the specifically religious dimension of early Daoist cosmology.

Modern scholars commonly understand references to the world in early Daoist writings, and particularly in the *Laozi*, as referring to the arts of rulership. Indeed, the *Laozi* often refers to the King as one who has responsibility for ruling and regulating the social world, but it also makes clear that the natural world, *tiandi* (Heaven and Earth), also referred to as nature (*ziran*), is ontologically very different from the social world, *ren* (the Human), also referred to as the empire (*tianxia*): the cosmological limits of the natural world extend far beyond the intentional control of any single individual wielding political authority. While the *Laozi* at times subordinates the social world to the natural world, the natural world itself commonly is subordinated to the pristine Dao. Although one can claim that there is an implicit, naturalist value judgment inherent in these subordinations, the *Laozi* is only working to establish a coherent model for presenting the distinctions between these three realms within the more encompassing and continuous cosmos. *Laozi* 23 offers a helpful clarification of these distinctions.

> Nature rarely speaks. Fierce winds do not last the whole morning; torrential rains do not last the whole day. Who makes these things? Heaven and Earth. If even (the realms of) Heaven and Earth cannot make these long lasting, how much more is this true for (the realm of) the Human? Therefore, one who takes up the project together with the Dao unites with the Dao. One who is of the *de* unites with the *de*.[19]

According to these views, not only is the natural world ontologically distinguished from the social world, it also possesses a range of presence hardly open to the average human. Nonetheless, it is clear that there is some kind of access for human beings, in some way involving the natural world, that allows for the possibility of uniting with or becoming one with the Dao and the *de*. This access lies squarely in "taking up the project" 從事 (*congshi*) that, as we have seen and will see again, refers to the Sage taking up the project of completing the world and incepting the harmony of Heaven-the Human-Earth. These and similar passages are often interpreted as demonstrating a strong political component; undeniably there is a political component, which is often made explicit in the writings themselves. Nevertheless, at virtually every turn we witness the almost ritualistic devaluation of the position of the King through the strict limitations ideally imposed on his ability to have any positive effect on the maintenance of order in the world. The tangential part played by rulers in the *Laozi's* vision of the harmonious world is underscored by their systematic relegation to thoroughly passive roles in terms that portray him as a simple figurehead. The duties of a King are, in the ideal formulations of the *Laozi*, negative in essence—he should do nothing that adversely disrupts the harmonious balance of the lives of his subjects, but act rather as a kind of babysitter. *Laozi* 75 states:

> People suffer famines because their superiors tax them to excess: for this they suffer famines.
> People are difficult to govern because their superiors pursue their own agendas: for this they are difficult to govern.
> People trivialize death lightly because their superiors are consumed with their own pursuit of life: for this they trivialize death.[20]

Laozi 53 states:

> The courts are spotless while the fields are full of weeds, and the granaries are all but empty. Their clothing is richly embroidered and colored, and they carry polished swords on their hilts. They eat and drink to excess, and have a wealth of possessions and goods. This is called thievery. And thievery is certainly not the Dao.[21]

The subsidiary nature of the political is brought out in even clearer terms in *Laozi* 32 and 37; here, the sphere of influence under the King's authority is juxtaposed with the creative potency of the pristine Dao, and pales in comparison. Outside of his duty to stay out of the way of other people and to try to influence other people to stay out of each other's way, his position does not appear particularly relevant for the effectuation of the world's harmony. *Laozi* 32 states,

> The Dao is constantly without name. Untouched it is small yet nobody in the world would dare to treat it as a subject. If marquises and kings were able to preserve it, then everything would spontaneously submit, and Heaven and Earth would come together and release sweet dew. Spontaneously (the sweet dew) would fall equally on everything, with no single person ordering that it be so.[22]

Laozi 37 also states,

> The Dao constantly performs non-intentional action yet nothing is left undone. If the Prince and the King were able to preserve it, everything would spontaneously transform. After having transformed, if their desires were once again to act up, I would subdue them with nameless simplicity. Nameless simplicity is to be without desire. To be without desire is to aim at the tranquility whereby Heaven and Earth will spontaneously align.[23]

In this last passage, the identity of the first-person speaker 吾 (*wu*) has been a point of contention: is this the King, the Dao, or the Sage? First, it cannot be the King; the King has already been mentioned, and if the "I" belonged to the King, then the use of the term "King" in the passage would be anomalous. Second, in later Daoist writings, the Dao is sometimes represented as speaking in the first person, particularly in the Tianshi writings; however, in early Daoist writings this usage is extremely rare. Finally, there are other sections of the *Laozi* where the Sage indeed does speak in the first person, and it seems likely that other chapters with an implied first person also probably presuppose the Sage. Further, the Sage, and the Sage alone, is invested with the capacity to transform the world actively through such means as simplicity and non-intentional action. If the first-person pronoun indeed does belong to the Sage, and the result of his nameless simplicity and non-intentional action effects spontaneous transformations, including the spontaneous alignment of Heaven and Earth, then the role of the King is severely compromised. *Laozi* 57 gives the clearest example of these several points: "Therefore the Sage says: 'I perform non-intentional action and the people are spontaneously transformed. I love tranquility and the people are spontaneously ordered. I pursue no affairs and the people spontaneously prosper. I have no desires and the people are spontaneously simple.' "[24] Clearly, it is the Sage speaking in the first person. The significant role fulfilled by the Sage, as evidenced in this as well as many other passages in the *Laozi*, stands in direct contrast with the thoroughly unimportant role played by the King.

The Sage is the true source of the ordering of the empire (*tianxia*), and the Sage is capable of this because he has gotten the Dao and merged with the world. Although the Sage does not control the world, the fact that he embodies the Dao in the world means that he realizes the cosmological possibility by which the realms of Heaven, Earth, and the Human unite in harmony. In this sense, the privileged representative of the realm of the Human is not the King but the Sage who exercises the influence of harmony that can unite the three realms.

The distinction between the (public) King as the political ruler and the (hidden) Sage as the genuine source of order is described in many different ways in the *Laozi*. *Laozi* 17 states: "The existence of the supremely high is not known. Below them, there are those who are loved and praised. Below them, there are those who are feared. Below them, there are those who are loathed."[25] Typical interpretations of this passage read the three figures mentioned as referring to three categories of the King, but the text itself does not literally make this designation. The text uses the term *shang* 上; *shang* has many meanings, but the exact referent is somewhat blurred in this instance. *Shang* can refer to one who is exalted, to the King, to superiors, or simply to the highest. In this instance, the supremely high 太上 (*taishang*) describes one who is hidden and unknown, but the King is eminently visible. It is interesting to note that the nominalization of *taishang* first appears in this passage of the *Laozi* to describe one who is hidden; later Daoist writings would appropriate this term and use it as the primary title for the deified Laozi, Taishang Laojun 太上老君. Neither Laozi nor Laojun were ever represented as the King.[26] Therefore, the supremely high most likely refers to the Sage, while the three persons mentioned below him refer to the good King who is praised, the bad King who is feared, and the despotic King who is loathed. Once again, the King is systematically subordinated to the Sage, and the Sage is hidden because he has merged with the world and its inhabitants. References to the combined qualities of the Sage as one who is hidden, unknown, and the true source of order, are extremely numerous. *Laozi* 22 states,

> Therefore, the Sage holds to the One and acts as the model for the empire. He has no vision of his own and thus he understands. He does not make himself seen and thus he is radiant. He does not battle with himself and thus is effective. He is not arrogant and thus can be long lasting. It is because he does not fight that nobody can fight him.[27]

The *Laozi* teaches the reader how to get the Dao and thereby become a true ruler, but the true ruler is consistently identified as the Sage, not the King; it is directed to the Sage, or more precisely to the potential Sage, and not to the King. The potential Sage can be, but is not necessarily, the King; he is any human being in general. The *Laozi* discusses individual human beings in one of three ways: as human being (*ren*), Sage (*shengren*), or King (*wang*). "Human being" signifies the Sage inherent as a potential in every human individual; "Sage" signifies the human being who has gotten, embodies, or possesses the Dao; and "King" is used in one of two ways: either it commonly refers to the public ruler who holds political office, or it rarely refers to the hidden ruler who holds no political office but is responsible for the harmonious relation between the social and natural worlds. This follows as a spontaneous consequence of there being one in the world who embodies the Dao.

The political tenor of *Laozi* 25 often serves as evidence for readings that see the text as a manual of statecraft, because this chapter supposedly reveals in crystal-clear fashion an ideological vision that accords a central position to the King, but these readings seem to miss the point. It says, "The Dao is great; Heaven is great; Earth is great; and the King is also great. In the center of the realms there are four greats, and the King occupies one place among them."[28] In light of my argument, one would expect the *Laozi* to name either the Human or the Sage in place of the King in order to set forth the strict triad of Heaven-Earth-the Human. It does not do this, however, because it is not here speaking of the cosmological structure of the world, but simply of the social and political world *tout court*. Indeed, its reference to the King is primarily employed to signify his figurehead position that, for all practical purposes, is essentially empty. This is evident because the portraitlike view presented by these lines is entirely static. This passage represents the commonsense view of the powers that be from the perspective of the social world bereft of any higher religious significance.

Reading these lines, one gets the feeling that there is a tremendous gulf standing between the common person, on the one side, and the King, Heaven, Earth, and the Dao on the other. The picture presented here is extremely political, constructed on hierarchical notions of superior/inferior; this is a pageantry display of the powers that the immediately following passages turn upside down. Situated in such a way that they bring out a powerfully contrasting perspective, they present a significantly different take on the true nature of the cosmological openness made available to human beings by the inherent structures of the world: "The Human models Earth; Earth models Heaven; Heaven models the Dao; the Dao models what is so of itself."[29] This passage radically undercuts the value of the position held by the King as the representative of the realm of the Human, asserted from the political vantage of the commonsense point of view. In contrast to the starkly static picture painted in the first passage, this passage represents the contours of a dynamic cosmological structure inherent in the very fabric of the world and available to any human being. Not only is this passage not specifically directed to the King, it also reasonably can be surmised that the King would experience particularly difficult hardships in pursuing the cosmological project. This is, again, a central component of early Daoist writings: the Sage is constantly at pains to avoid the many offers of the King who would attempt to put him directly on the throne, precisely because the figurehead duties and responsibilities will necessarily hinder the cosmological project of completing the harmony of the world; the Sage always spurns the King.[30] In early Confucian discourse, on the other hand, the Sage reluctantly does come to accept the King's desire for his enthronement; the best examples of this are Shun, who accepted the throne from King Yao, and Yu, who accepted the throne from King Shun. In early Daoist discourse, the enthronement of the Sage is unthinkable, and the *Zhuangzi* provides two interesting illustrations of this. The first comes from *Zhuangzi* 1; here the King is Yao, and the Sage is Xu You.

> Yao tried to abdicate the empire to Xu You. He said, "If a torch is not extinguished after the sun and moon come up, isn't it difficult to see its fire? If the channels are still being flooded after the seasonal rains begin to fall, isn't this too much labor in working the fields? As you stand, the empire is ordered but I still hold the seat of honor. In my eyes, I do not deserve it, and I ask you to govern the empire." Xu You said, "You are the one who orders the empire. If the empire has already been ordered and then I take your place, wouldn't it be only for the sake of the name? Names are only the guests of the true substance. If I take your place, wouldn't I only be playing the part of the guest? . . . Return to your place, lord, I am of no use for serving the empire."[31]

Zhuangzi 17 states:

> Zhuangzi was fishing in the waters of Pu. The King of Chu dispatched two officials to deliver this message to him: "It is my desire to burden you with the realm." Zhuangzi continued to hold his fishing rod and without turning his head, said, "I have heard that there is a sacred tortoise in the state of Chu, and that it has been dead for three thousand years. The King keeps it wrapped in a chest on the top of his royal temple. Now, do you think that this tortoise would rather be dead and have its bones venerated, or would it rather be alive and drag its tail in the mud?" The two officials said, "It would rather be alive and drag its tail in the mud." Zhuangzi said, "Then be gone! I am going to drag my tail in the mud."[32]

These *Zhuangzi* passages are indicative of early Daoist attitudes concerning the reasons why the Sage invariably rejects the offers of the King, in which even the King realizes the emptiness of his position in comparison with the Sage, and these attitudes represent another defining element of early Daoist discourse.

This last section of *Laozi* 25 is significant also for a related set of issues: this is the single passage in which the *Laozi* directly mentions the strict triad of Heaven-Earth-the Human and its relationship with the Dao. What we see almost appears as a sequential ordering of the hierarchical ranks of cosmo-political authority, which would appear to mitigate against my reading of the cosmological possibilities involving the triad for which I have been arguing. However, that interpretation would be more likely if this sequence started with the Dao, went through Heaven and Earth, and ended with the Human, thus following typical hierarchical orderings that prioritize first terms and subordinate second terms, as seen in the contrasting first passage of *Laozi* 25; on the contrary, this passage presents a soteriological path of reversal leading from the Human to the Dao, in which Heaven and Earth can be seen as holding a position midway between them, and in which everything ultimately will result in a state of being "so of itself." One further mention of the King is telling in this regard; *Laozi* 16 states,

> Extending emptiness to the limit and preserving tranquility in the center: the ten thousand things come bursting forth and by this I see their return. Everything flourishes and flourishes, and each one returns again to its root: this is called tranquility. Tranquility is returning to your fate. Returning to your fate is said to be constancy. Knowing constancy is to be illuminated; not knowing constancy is to act blindly for disaster. Knowing constancy is to be inclusive. Being inclusive leads to impartiality. Impartiality leads to Kingship. Kingship leads to Heaven. Heaven leads to the Dao. The Dao leads to longevity, and the body suffers no harm.[33]

In this passage, this soteriological process of reversal and of returning to the Dao is set forth very clearly. Although it mentions Kingship, this passage in no way describes a political reality based on imperial authority. In what follows, I will argue that the ultimate consequence of becoming one with the Dao concerns a physical benefit for the body and not the state, a conclusion already stated in the *Taiyi Sheng Shui*, and one that will be found time and again throughout this course of this study.

The *Laozi* tends to describe the triadic union of Heaven, Earth, and the Sage, with the passage just discussed representing an important exception; other early Daoist writings will commonly speak of the triad in terms of Heaven, Earth, and the Human. The Human does not refer to human beings, and the realms of Heaven, Earth, and the Human precede the existence of human beings. The ultimate harmony of the realms of Heaven, Earth, and the Human remains incomplete until a human being becomes a Sage and physically manifests the Dao through embodiment, effectuating their ultimate harmony and bringing the project to completion. This completion itself is best understood as a second-order, restored harmony. These are central ideas concerning the cosmological completion of the world and the soteriological possibilities of human beings in it, because the world always represents the space from which the soteriological progress is initiated. For example, *Laozi* 29 states, "As for one who desires to take the world by acting on it, I see that such a one will never obtain it. The world is a sacred vessel; it cannot be acted on and it cannot be held. One who acts on it loses it, and one who grips it displaces it."[34] Along the same lines, *Laozi* 5 says: "Heaven and Earth are not humane; they regard the ten thousand beings as straw dogs. The Sage is not humane; he regards the common people as straw dogs."[35] Straw dogs were used as ritual substitutes within the early Chinese sacrificial environment. Both passages portray the world itself as the sacred and ritual arena in which the Sage achieves unification with the Dao.

From the time when the world first comes into being and ever after, it subsists as the general ground for the continuous presence of the pristine Dao. The Dao and the world share a particular kind of relationship; while the Dao cannot be identified simply as the world, nor the world identified simply as the Dao, they are also not ontologically distinct, and the precise relation between the Dao and the world is a primary object of meditation in all Daoist writings. The relation is commonly characterized in terms of seeing the Dao as a mother and the world as the child receiving constant nurture. These images are so powerful both in early and later Daoist writings in large part because the world is never seen as a finished product, created once and for all, but on the contrary was envisioned as a work constantly in progress, like the perpetually young child requiring a mother's nurture. *Laozi* 51 qualifies these images of birth and nurturance by stating that the Dao does not take possession of that to which it gives birth; while these images appear throughout the *Laozi* in descriptions of the activities of the Dao, they often also describe the activities of the Sage in hiding. "Thus the Dao gives birth to them and the *de* nurtures them. It raises them and rears them, fosters them and nurses them, feeds them and shelters them. It gives birth to them but does not possess them; provides for them does not command them, raises them but does not control them. This is called profound *de*."[36] The relation between the world and the Dao, then, is like the relation between a mother and her child.

Heaven, Earth, and the Sage are "not humane" (*bu ren*), claims *Laozi* 5, not because they are heartless and uncaring, but rather because their total system of relations represents the sacred context for early Daoist cosmology. Heaven, Earth, and the Sage are not humane because they do not act directly or intentionally in providing the kind of humane attention that comes from a superior and is directed to an inferior. The world, as *Laozi* 29 says, is a sacred vessel; it cannot be dominated, subordinated, or transcended. Directly acting on the world without having the Dao embodied, the typical behavior of the typical King, profanes its sacred nature, and in doing so one immediately loses the world. Yet when there is a Sage who embodies the Dao present in the world, it undergoes a powerful transformation. Although the Sage performs non-intentional action (*wei wu wei*), pursues no affairs (*wei wu shi*), and embraces tranquility, it is due solely to his radiating influence, mirroring the mysterious *de* 玄德 (*xuande*), by which the world is brought to completion. This completion spoken of by the *Laozi* refers to the completion of the world's sacred potentiality inherent from the moment of its initial emergence in the cosmology. To complete the world is to complete the cosmology, and the Sage resides in the very core of this project.

The Sage embodies the Dao in the world, and he completes the world by merging with it, thereby setting the Dao free within the three realms. *Laozi* 49 describes this merging: "The Sage has no constant mind, he takes the mind of the common people as his mind. . . . When the Sage is present in the empire (*tianxia*) he is one with it, and he merges his mind with the empire (*tianxia*)."[37] Here, "the empire" refers to the realm of the Human. The strictly religious context of the relations among Heaven, Earth, and the Sage, standing as the supreme representative of the realm of the Human, is clearly set forth in *Laozi* 7: "Heaven endures and Earth is long-lasting. The reason why Heaven can endure and Earth can last long is that they do not live for themselves; therefore they can live long. For this reason the Sage puts his body last but his body is first; he regards his body as something distantly related to him but his body is preserved."[38] The completion of the ultimate harmony of the triadic world is marked by the Sage's merging with it. The *Laozi*'s earlier denial of humaneness to Heaven, Earth, and the Sage underscores the spontaneous, nondominating process of the world's ultimate harmony.

Zhuangzi 25 further pursues this notion of the ability of the Sage to merge his mind with the empire and thereby initiate a great harmony.

> The Sage attains such intimate union [*choumou* 綢繆 denotes sexual union] with things as that of a single body, for it is only his nature.

> . . . Thus the Sage causes his family to forget their hardships when they are impoverished, he causes the Duke and the King to forget their titles and emoluments when they are victorious, and transforms them into humble people. When he is among things, he takes pleasure with them; when among people, he experiences their pleasures but preserves himself. Therefore, he sometimes says nothing and yet he immerses people in harmony. When he stands together with others, he causes them to be transformed.[39]

Beginning with the *Laozi*, the vision of the world consisting of the three realms of Heaven, Earth, and the Human became the standard model for all formulations of the cosmological world in early Daoist discourse. Its establishment played a foundational role in the development of early Daoist cosmology primarily by its ability to set forth a clear working model of the world in which humans were seen capable of achieving union with the pristine Dao. As I will argue next, this model made possible the clearly articulated visions of the transformative, soteriological potential accorded to both human beings and the world that lie at the very core of the *Zhuangzi* and the *Huainanzi*.

WHY POLITICS AND RELIGION DON'T MIX; OR DO THEY?

The cosmological model of the world adopted in early Daoist writings, conceived in terms of the three realms of Heaven-the Human-Earth, accommodates the position of the King by giving him a figurehead position. Relegating the King to such a position poses an ideologically radical, though physically nonthreatening, challenge to his authority that the several non-early Daoist traditions tend to ignore. In the eyes of the early Daoist writers, true authority to order and complete the world always lay with the Sage. As this early Daoist cosmology spread throughout the world of early China, other traditions began to adopt its general structure and parameters, while at the same time investing a much more substantial authority to the King than the early Daoist writings were willing to allow.

Historically, it appears that the writings often identified with the tradition of Huang-Lao Daoism, with its sociological roots in the Jixia Academy in the state of Qi, played a transitional role here.[40] They brought the early Daoist cosmology into alignment with the techniques of rulership, and this marks the defining feature of this tradition. These Huang-Lao writings still attribute to the Sage full responsibility for completing the world, but they reveal a distinct tendency to associate the Sage and the King to a degree not seen in the mainstream writings of early Daoism. The primary writings for this tradition, the *Huangdi Sijing* 黃帝四經, still maintain the early Daoist emphasis on the Sage, but they inject sophisticated techniques of rulership into the cosmological model of the three realms of Heaven, the Human, and Earth. In these writings, Huangdi, the mythological first ruler, fulfills the role of both the Sage and the King.[41] Huang-Lao Daoism appears to have taken as one its fundamental pillars the exploration and exploitation of the political consequences of the presence of the Sage in the world as given in the *Laozi* text, a consequence that other early Daoist writings do not fully explore. The cosmological structure of the world viewed in terms of the three realms of Heaven-the Human-Earth became a standard trope for Chinese literary practices through such early employments as are found in the writings of the *Huangdi Sijing*, and most certainly influenced the Confucian readings of early Daoist discourse.

The *Huangdi Sijing* is in five parts; these are *Jingfa* 經法, *Jing* 經, *Cheng* 稱, *Daoyuan* 道原, and *Jiu Zhu* 九主. Section four of the *Jing*, *Liu Fen* 六分, demonstrates an obvious politicization of the three realm cosmology by affirming that the substantive authority of the King is the natural consequence of his power to represent the third term of the triad, the Human, and thereby complete the world. In traditional early Chinese rhetoric, Heaven is said to cover, and Earth is said to uphold; here we see the true ruler appropriate these two powers by virtue of his ability to participate in the full triadic mechanism. Later in the same section, it says that the King, by relying on the Dao of the true ruler, can participate in the powers of the three realms, and only by so doing can he be a true King.

> When the world is at Great Peace (*taiping*), he rules with bright potency and makes a triad (*can*) with Heaven and Earth that is able simultaneously to cover and to uphold. Without entertaining private bias, he is therefore able to rule the world as a true King. The Dao of he who rules the world as a true King has Heaven in it, has Earth in it, and has the Human in it. Because he participates with them and uses them equally, he rules as a true King and possesses the world.[42]

The fifth section of the *Jingfa*, *Sidu* 四度, pursues this theme of the true ruler's forming a triad with Heaven and Earth in the accomplishment of the perfect world:

> Balancing movement and quiescence and forming a triad with Heaven and Earth is called "civility;" to punish . . . in a timely fashion is called "martiality". . . . Forming a triad with Heaven and Earth and harmonizing with the hearts of the people, while concurrently establishing civility and martiality, is called "the Supreme Union" (*shangtong*).[43]

These writings manifestly demonstrate their political commitment to upholding the position and authority of the true ruler who individually represents the realm of the Human by forming the third term of the triad together with Heaven and Earth. When this occurs, it is described as the achievement of "Great Peace" 太平 (*taiping*) or "Supreme Union"

上通 (*shangtong*). These and similar phrases are used throughout these writings to represent the completion of the world expressed in images of the triad. They consistently signify the active agent of this completion with the King, not the Sage, and this appears to represent a radical departure from the thrust of the early Daoist writings. Nevertheless, what clearly has not changed is the notion that the world as it is given to normal experience, imaged in terms of the model of Heaven-the Human-Earth, remains an unfinished project, and the responsibility for completing the world continues to lie with the true ruler as the exemplary representative of the realm of the Human as the third term.

The second text of the *Huangdi Sijing*, the *Jing*, in many ways appears to be far less willing to honor the cosmological role of the King in the explicit terms adopted by the *Jingfa*, and it more closely coheres to the early Daoist discourse; it consistently identifies the Sage as the major representative of the realm of the Human. It should be noted that this chapter as a whole is dominated by the myths of Huangdi that depict his fashioning of the phenomenal world. Huangdi himself plays an extremely ambivalent role in the writings of early China, and he was represented, at various times in the various cycles of the myths about him, either as the King who founded kingship, or the Sage who attained immortality. These different representations are at times very difficult to distinguish; in the *Jing*, they often overlap. Given that all of the writings of this chapter center around Huangdi as either King or Sage, it is impossible to state definitively whether the conceptions of the completion of the world in this chapter are best interpreted as taking the Sage or the King as the exemplary representative of the realm of the Human. In any case, the *Jing* vigorously adopts the cosmological model first set forth in the early Daoist writings. In the first section, *Li Ming* 立命, Huangdi states, "I received the mandate from Heaven, established the position on Earth, and gained my reputation from the Human."[44] Section ten, *San Jin* 三禁, presents a detailed instance of the differentiated, limited, yet interactive domains of the three realms.

> According to the prohibitions of Earth, one is not to take away from the high, nor add to the low. Do not block up the rivers; do not oppose agricultural tasks; do not oppose the people's high intelligence. . . . The Dao of the Human is both hard and soft. If it is too hard, it is not sufficient to be used. If too soft, it is not sufficient to be relied on. . . . The Dao of Heaven is distant, yet it spreads to the land below and is applied to the Nine Regions. . . . Heaven has the sun that is constant.[45]

For early Daoist writings in general, and this piece in particular, the term Heaven signifies the celestial realm, Earth signifies the geographic world, and the Human signifies the cultural sphere. These writings bear witness to a fully developed understanding of the differentiated realms of Heaven-the Human-Earth and their associated domains of signification. In order to examine the more precise understanding of these realms, we will be best served by taking account of the mythological portions of this chapter. Those portions give vivid firsthand depictions of Huangdi's success in bringing each of the three realms to a cosmological completion.

The original, cosmological event of the initial emergence of the three realms achieved only a temporary form of completion, which I call the first-order completion. By calling the initial cosmological establishment of the world, which owed a great deal to the activities of Huangdi, a first-order completion, I mean to say that its formation participated in the natural unfolding of the stages of the cosmological sequences. I will shortly speak of the cosmological completion, effectuated by the Sage according to early Daoist discourse, and the King according to non-Daoist discourse, as the second-order completion. The *Jing* is remarkable in that it gives precise and vivid depictions of both completions. The clearest instance of the second-order completion as given in the *Jing* unmistakably identifies the Sage as the agent who completes the world; further, this passage also unmistakably uses the identical term of the *Laozi* to refer to this completion, namely, *shi* 事, "project." This alone seems sufficient reason to identify this chapter as lying squarely within the domain of early Daoist discourse. This passage starts by describing the specific project of Heaven, referring to the proper movement of the seasons; it then describes the specific project of Earth, referring to the timely production of the agricultural harvests; and finishes by describing the specific project of the Sage; it says: "When the Dao of Heaven has been carried out, Earthly things will then be prepared. Mutually completing the scattered and flowing is the project of the Sage."[46] This passage gives a very clear account of what is entailed by the notion of completing the project.

Before leaving the *Huangdi Sijing*, I want to comment on its use of *yin-yang*, and the dramatic possibilities for vastly expanding the cosmological field of signification implied in it. Although *yin-yang* is only briefly mentioned in the *Laozi*, the *Sijing* constantly correlates a tremendous variety of things to them. The most striking instance of this comes in the third chapter, *Cheng*: "Heaven is *yang* and Earth is *yin*. Spring is *yang* and autumn is *yin*; summer is *yang* and winter is *yin* [There follows twenty more correlations of objects and phenomena with *yin-yang*] . . . All that is *yang* is modeled on Heaven. . . . All that is *yin* is modeled on Earth."[47] The application of *yin-yang* to the cosmological model of the world also appears throughout the *Huangdi Sijing*, where it represents an additional set of notions that greatly assist the full development of the cosmology. The *Cheng*, which again seems to be more aligned with the political commitment revealed in the *Jingfa* by its use of the King rather than the Sage, also says, "Therefore one who is a true King does not govern the state by means of luck; he governs the state by steadfastly putting the Dao first. Above, he knows the seasons of Heaven. Below, he knows the benefits of Earth. In the middle, he knows the affairs of the Human. He is good at *yin-yang*."[48] This tendency of calling upon *yin-yang* is also evident in the early Daoist writings; the expansion of their field of signification in their discussions of the cosmological completion greatly benefited from the addition of the set of notions associated with *yin-yang*, and this is particularly evident in their continued commitment to taking the Sage as the primary agent for that completion.

A further consequence of the establishment of this cosmology is that it served as one of the great transitional bridges from the early Daoism of the Warring States and early Han to the institutionalized Daoism first established in the late Han. Although the writings of later Daoism certainly tampered with this model, they never decisively departed from its general outlines. Moreover, this cosmology was adopted by all other literary traditions of China both early and late, even to the point at which it became representative of "Chinese" cosmology in general. Particularly in the hands of the Confucian writers, this model, slightly adapted to fit a King-centered rather than a Sage-centered empire, became standard rhetoric. This is clearly seen in the works attributed to Xunzi in the Warring States period, and Dong Zhongshu in the Han period.

In the terms of their own discourse, it was the Confucians who successfully undermined the position of authority held by the Sage, and inserted the King in his stead. The *Xunzi* was the first Confucian writing to demonstrate a clear awareness of this strategy and complete the substitution of the King for the Sage once and for all with respect to non-Daoist writings. The essay, *Tian Lun* 天論, stands as the primary source for the later Han period adoption of this model when the Confucians institutionalized their hold on the bureaucratic system of control. Xunzi, who was certainly exposed to this model of the world during his lengthy stays at the Jixia Academy, even appears to cite the *Huangdi Sijing* in his essay. A primary consequence of this essay was that it strictly divorces the Sage, as the highest representative of the realm of the Human, from the mechanisms whereby the world comes to completion; further, it seems to condemn all those who would even consider reinstating the Sage to the role of completing the world in the terms of the early Daoist discourse. Differing greatly from the Sage of the *Laozi*, the Sage of the *Xunzi* consciously refrains from any direct interaction with the realms of Heaven and Earth, which, instead of forming a single harmony with the Human, will ever remain contained in their own specific realms. For example, the *Xunzi* says:

> To complete without acting, and to obtain without seeking, indeed may be described as the task of Heaven. In such a situation, the [perfect] human, however profound, does not add his own thought to it; however great, he does not add his own abilities to it; and however refined, he does not add his own scrutiny to it. This is called "not competing with Heaven in its task." "Heaven has its seasons; Earth has its resources; and the Human has its government." This is called "being able to form a triad." When a human abandons what

> he should use to form the triad, yet he longs for the benefits that result from the triad, this is confusion indeed.[49]

These kinds of Confucian writings have played a dominant role in shaping modern readings of the cosmological model of the world as found in early Daoist discourse. Moreover, inspired by these Confucian appropriations, this Confucian-inspired interpretative overlay has been decisive in allowing modern scholars to claim that the *Laozi* is a work of political philosophy while overlooking its specifically religious dimensions.

THE WORLD WAS BORN, NOT MADE

Early Daoist discourse is most immediately identifiable by its constant dependence on notions used to articulate the pristine Dao, and the *Laozi* stands as its first and primary representative. Appearing in the initial stages of the emergence of that discourse, it decisively influenced all other early Daoist writings, as well as the religious, political, and literary sensibilities of early and traditional China more generally. Our modern understanding of the role played by the *Laozi* in early Daoist discourse is better informed than ever before, owing to the fortuitous discoveries of the Guodian and the Mawangdui *Laozi* bamboo slips. The early exercise of its influence appears most clearly in the two major collections of early Daoist writings, the *Zhuangzi* and the *Huainanzi*; both commonly refer to Laozi the Sage or directly quote passages from the *Laozi*. On a less tangible measure, both of these two works consistently rely on images, metaphors, and notions that appear for the first time in the *Laozi*. Demonstrating that the internal contents of the *Laozi* either had or had not been in circulation before the appearance of the earliest datable documented evidence for it (now provided by the Guodian slips) is significantly less important than understanding that this short text, with its terse and in many ways minimalist discourse, for the first time cements an entire range of images, metaphors, and notions that become standard elements of the religious, philosophical, and literary sensibilities of early Daoism and beyond.

The deep influence of the *Laozi* over all subsequent early Daoist writings is primarily seen in their wholesale reliance on the *Laozi*'s formulations of the pristine Dao and the cosmology and soteriology through which those formulations are presented. The important points of the cosmology concern the triadic model of the world, represented in terms of Heaven, the Human, and Earth, that is born from the body of the pristine Dao. The important points of the soteriology concern the central role of the Sage who embodies the pristine Dao, completes the world, and enjoys a longevous body. Due to the terse and minimalist nature of the *Laozi*, the model of the triadic cosmology is explicitly indicated only in a very few instances, yet it nonetheless supplies the enduring structure through which are articulated the soteriological possibilities for human beings existing in the world.

The minimalist nature of the *Laozi* is most obvious in *Laozi* 42. The conciseness of this passage is remarkable, and stands in direct contrast to the powerful and decisive influence it exercised over all subsequent cosmological visions: "Dao gave birth to the One. The One gave birth to the Two. The Two gave birth to the Three. And the Three gave birth to the ten thousand things. The ten thousand things carry *yin* on their backs and embrace *yang*. Through the blending of *qi* they arrive at a state of harmony."[50] This short passage sets forth the most influential portrayal of the supremely Daoist vision of the stages of the cosmology through its depiction of the sequence of the original emergence of the world from the pristine Dao. The emergence of the world culminates in the existence of the ten thousand things, otherwise known as phenomenal reality, and the resulting condition of the world manifests in an initial state of the first-order harmony of the cosmology. This initial harmony, however, could not endure because of the disruptions to it brought on by human interference, causing a substantial loss of the full presence of the pristine Dao in the world. But because the presence of the pristine Dao continues to remain in the world, albeit in a less than perfect way, the door is left open for the transformative second-order harmony of the soteriology, which is consistently described as a futural event. The standard sequence of other early Daoist cosmological visions of the first-order harmony invariably follow, with only slight modification, the sequential order presented in *Laozi* 42.

Laozi 42 appears as the earliest documented formulation of this Dao-based cosmological sequence in four stages, leading from to one to two to three to ten thousand. The *Laozi* does not make its numerical identifications explicit, and different commentators throughout the long history of *Laozi* exegesis interpret them in various ways. The most influential of the traditional Daoist interpretations (exemplified in the *Heshang Gong* commentary that informs my interpretation of *Laozi* 42) identifies the One with *qi*, the Two with *yin-yang*, and the Three with Heaven-the Human-Earth, with the ten thousand things referring to all phenomenal beings.[51] Later Daoist writings do not alter the numerological sequence, but they do extrapolate on it, most often by providing their own images of the environments and processes in order to flesh out the minimalist rhetoric of this passage.

Relying only on selected passages from other sections of the *Laozi*, it is possible to piece together a composite picture of the cosmological first-order harmony on which *Laozi* 42 is structured. In those other sections are found fuller descriptions of the actual environments and processes of the cosmological sequence. It begins with the pristine Dao, alone within the infinite reaches of empty nothingness, before the existence of time and space. In some sense of the term, the Dao is alive, and moves. The movement of the Dao is chaotic and without regulation within its own fields of that-which-is (*you*) and that-which-is-not (*wu*). The movement of the Dao into those areas of that-which-is bring with it its own presence, while the movement of the Dao away from those areas of that-which-is-not in which the Dao previously had been are left vacant, filling those vacated areas with its own absence. Gradually, these continuous movements of the Dao create a harmonious vacuum that, after countless passages of eons, fell into a simple rhythm of coordinated movement. This rhythm manifests itself as a wind or generates a wind that yet simply consists of the Dao in the dual modes of that-which-is and that-which-is-not. This cosmic wind is the breath of the Dao, identified by the term *qi*. This *qi* is hot and vaporous, much like the breath of a human, but teeming with the ingredients of all life. This is an environment of chaotic potential out of which all life physically emerges. This stage of the cosmological sequence corresponds to the One, and can be identified with the abyssal and placental waters discussed in chapter two.

This cosmic *qi*, again over great periods of time, gradually falls into a rhythm whereby its chaotic surgings, described for example by the Nine Abysses of the *Zhuangzi* and the *Liezi*, develop into oceanlike tides of expansion and contraction, exhalation and inhalation. The two breaths consolidate in spaces above and below, resulting in a nebulous differentiation. *Yang* is the upwardly consolidated breath, and *yin* the downwardly consolidated breath; together, these breaths correspond to the Two. These *yin-yang* breaths countless times separate and again unite in a kind of cosmic dance that is regularly depicted in strongly sexual terms and images underscoring their sexual potency for the generation of new life. The background for these images can be traced to the ancient and early myths, and directly calls upon the images and symbols associated with the primordial couple and their incestuous unions. These *yin*-inhalations and *yang*-exhalations describe the developed movements of the Dao: *yin-yang* together manifest the expansive and contractive movements of the Dao in general. Again over vast expanses of time, the gradual and relentless rhythm of their movements cause the breaths to establish their individual properties, thus making them differentiate to an even further degree. The exhalation breath of expansive *yang* is light and ascending, and the inhalation breath of contractive *yin* is heavy and descending. This process is commonly described in terms of muddy water coming to settle, whereby the clear and the turbid gradually separate.

The ascending *yang* congeals in the formation of Heaven, and the descending *yin* congeals in the formation of Earth. The congealing and formation of the two realms of pure *yin* below and pure *yang* above are distinguishable in theory, but because there is no absolute separation in the space between them, the upper parts of the pure *yin* and the lower parts of the pure *yang* continue to intermingle, forming in turn a realm standing between. This realm also is not empty but rather consists of a mixture of more or less equal parts of *yin-yang*; the *Laozi* describes this space as a cosmic bellows, an image that, as Harper shows, played a major role in the macrobiotic practices of early China.[52] *Laozi* 5 states: "The space between Heaven and Earth—is it not like a bellows? It is empty and yet not depleted. When it moves, more always comes out."[53] This middle space congeals in the formation of the realm of the Human. This middle realm is the direct offspring of the unions of *yin-yang*, also describable as the offspring of Heaven and Earth; either way, the underlying idea is identical. The world as a whole comprises the environments and processes that take form in this stage of the cosmological sequence and consist of the realms of Heaven-the Human-Earth, and this corresponds to the Three.

At this stage of the cosmology, the realms of Heaven-the Human-Earth unite in a state of original harmony, and all phenomenally existing things are brought to life from this union; this corresponds to "the ten thousand things." All existing things inherently possess all of the active elements of the cosmological sequence, including the Dao, *qi*, *yin-yang*, and even Heaven, Earth, and the Human. Heaven and Earth are directly included among the cosmological elements forming the constitution of whatever exists because the physical form is the gift of Earth, while the breath or spirit is the gift of Heaven. Here I give the final lines of the cosmology of the *Laozi* 42: "The ten thousand things carry *yin* on their back and embrace *yang*. Through the blending of *qi* they arrive at a state of harmony." The abstract image of this passage points to the birth of the infant, who ideally is birthed face up facing Heaven, thus "embracing *yang*," while the Earth holds up its back, thus "carrying *yin*." The infant's body exists in space in between the pure *yang* of Heaven and the pure *yin* of Earth, where the *qi* of *yin-yang* blends in harmony. The preponderance of *yin* or *yang* in any individual being or thing is a defining feature of the makeup and constitution of all existing things in the world; for example, *yin* predominates in fish and stones, while *yang* predominates in birds and stars. Human beings are a particular species among the ten thousand things, because they are in essence equal parts *yin* and *yang* at the moment of conception (gender differences result from a later predominance of either *yin* for females or *yang* for males during the period of gestation). Human beings also possess the further distinguishing qualities of the realm of the Human, including language, cultural production and reproduction, and intentional consciousness with its habit of distinction and judgment.

Humans first come into existence only in the final stage of the cosmological sequence, as one member among the ten thousand things. The genetic constitution of humans is equal parts *yin-yang*, originally, inherently, and ideally harmonized in the physical being of all humans, but especially the infant. This supreme harmony can be understood in both of two ways: first, as the natural constitution of humans who lived in distant antiquity, when the ten thousand things were first brought to life and all things were one with the presence of the Dao; second, as the natural constitution of the newborn infant who has yet to undergo the processes of socialization. The original, first-order harmony enjoyed by the infant is invariably lost at a later time; however, a fully grown human can, through different programs of physical cultivation, often called the arts of the Dao 道術 (*daoshu*), attain a restored, second-order harmony by properly circulating and rejuvenating the inborn cosmological elements, including *qi* and *yin-yang*; success at this results in the generation of the Sage.

The first-order cosmological sequence of *Laozi* 42, leading from the One to the ten thousand things, squarely locates the initial existence of humans at the final stage of the cosmic birthing process. This forward-moving sequence represents the unfolding of the different stages of the cosmos and the world; the reverse order sequence, from the ten thousand things back to the One, represents the second-order soteriological sequence, leading from phenomenal reality to the pristine Dao. Moreover, the first-order cosmological sequence is a process of externalizing the inherent materials and energies internal to each of the named elements (Dao, *qi*, *yin-yang*, etc.), and this is "to give birth to" (*sheng*); the second-order soteriological sequence is a process of internalizing these same named elements within the physical body of the human being pursuing the path to becoming a Sage. The Sage begins the soteriological sequence from the stage of the ten thousand things by initiating a reversal of the temporal and spatial limits of the world. This in turn leads the Sage to merge with, through internalization, the Three, understood as Heaven-the Human-Earth. At this stage, the Sage confronts and embodies the pure potency of the Two, understood as *yin-yang* in isolation from the celestial, natural, and social realms of the world. Embodying vital *yin-yang* by internalization, the Sage circulates them throughout the body, whereby they come together and unite countless times. The unions of *yin-yang* cause the revitalization of their separate potencies and because of this their unions are sustained. Sustaining their unions, they merge and revert back to their original unity in the One, understood as *qi*. This *qi* is the cosmic vapor, breath, and trace of the pristine Dao. At this stage, the Sage isolates and embodies original *qi* in its pure form before its separation into *yin-yang*. Embodied, original *qi* manifests as the pristine Dao to which the Sage then unites. This is, essentially, the second-order soteriological sequence, structurally the reverse sequence of the first-order cosmology, and early Daoist writings regularly designate this as "getting the Dao" 得道 (*dedao*) or "embodying the Dao" 體道 (*tidao*).

SAGES LIVE THE ADVENTURE

The cosmological sequence culminates in the first-order harmony of the world, but the world continues to exist as continuous potential that is ever only partially realized. The soteriological sequence culminates in the second-order harmony of the world, and only through this is the potential of the world fully realized. The Sage serves as its primary agent, and he is able to accomplish this because he embodies the pristine Dao in his very body in the midst of this very world. Because he embodies the Dao and successfully completes the reverse-order soteriological sequence, the Sage completely merges with all parts of the world and cosmos; mastering the world in this way, he attains a complete freedom that accrues from his unity with the Dao. The Sage enjoys a total access throughout the world and the cosmos.

Although early Confucian discourse speaks about many Daos (the Dao of the Gentleman, the Dao of the Sage, the Dao of King Wen, etc.), they tend to emphasize two closely related Daos: the Dao of Heaven 天道 (*tiandao*) designates the supreme moral authority of Heaven, and is identified with the true King who carries out the will of this moral authority; and the Dao of the Human 人道 (*rendao*) designates the ethical behavior of all humans below the King, best revealed in the proper exercise of filial piety 孝 (*xiao*) with its sociopolitical ramifications. Early Daoist discourse focuses on three distinct Daos: the pristine Dao plain and simple, often called the Dao of antiquity 古之道 (*guzhidao*); the Dao of Heaven, designating the natural world, which is consistently exalted; and the Dao of the Human, also called the Dao of the Sage 聖人之道 (*shengrenzhidao*), which is sometimes exalted and at other times denigrated.[54] The first Dao originates as virtually the exclusive possession of early Daoist discourse; indeed, it seems to be the initial inspiration for this discourse. Early Daoist employments of the second and third Daos are used in ways that are radically different from those of the early Confucian discourse, though they share the same name. In the *Laozi*, the Dao of Heaven and the Dao of the Human/ the Dao of the Sage signify two different realms of existence, whereas in the *Zhuangzi* they signify the natural and the artificial in the developed constitution of the matured human. I will examine these employments of the *Zhuangzi* in more detail in a later chapter. *Laozi* 81 states: "Thus the Dao of Heaven is to benefit and not to injure; the Dao of Sages is to act on behalf of others and not to be competitive."[55] These usages reveal the redeeming difference between the realms of culture and nature, underscoring the critical separation between human social existence, deemed to have lost its natural qualities, and the spontaneous existence of the Dao. *Laozi* 77 brings out the veritable distance separating these two Daos:

> Is not the Dao of Heaven like the flexing of a bow? It presses down the high and raises up the low; where there is a surplus it takes away, and where there is a deficiency it gives more. The Dao of Heaven is to reduce the excessive and increase the insufficient; the Dao of the Human is not the same. Where there is a deficiency it takes away, and where there is a surplus it gives more. Now, who is able to have a surplus and use it to give to the world? Clearly, only one who possesses the Dao. Like this the Sage takes action but does not hold onto (the results), and he makes merit but does not claim credit. Like this is his desire not to make a display of his worthiness.[56]

The phrase, "where there is a surplus it takes away, and where there is a deficiency it gives more," describes the natural, life-sustaining mechanisms inherent in the natural world, which nurtures and supports all beings equally. The phrase, "where there is a deficiency it takes away, and where there is a surplus it gives more," describes the unnatural, life-destroying practices of rulers and the social elite, who take what little the common people have to depend on for their sustenance in order to finance their own desires. These images make explicit the growing separation of the realms of nature and society even to the point that they become ontologically distinct, with the consequence that the full spontaneity and vitality of the pristine Dao is shut out of the realm of the Human. In essence, this describes yet another stage of the cosmology, in which the world as a whole, made up of the realms of Heaven-the Human-Earth, themselves separate into two distinct realms, the realm of Nature and the realm of Culture. This separation occurs with the breakdown of the first-order harmony as the realm of the Human loses its oneness with the spontaneous, natural movements of the pristine Dao. Now, it is only the Sage who can break through the confines of Culture and break into the realm of Nature; by doing so, he overcomes the separation and allows it to heal. The *Laozi* notably brings in the notion of the completion of the project at this point; the ability of the Sage to break through these realms in order to reunite them is a central component for the cosmological completion the world. From within the confines of the Dao of the Human, the Sage alone wins complete access to the Dao of Heaven. Doing so, he effectuates the first steps of the process of the project of completing the world. The reason for this ability of the Sage to harmonize is that he alone has achieved possession of the pristine Dao by returning to the source of all life where there are no distinctions of any kind. Embodying the pristine Dao, it is thereby made present in the world, and this instates the second-order harmony.

This separation between the natural world of Heaven and Earth and the cultural world of the Human can only be healed by a soteriological return to the full presence of the pristine Dao. Successfully accomplishing the soteriological return means that the Sage breaks through all of the stages of the cosmological sequence, while continuing to remain in this world. The Sage embodies the Dao, and travels on far-off journeys back and forth through the stages of the cosmology and the soteriology: he enjoys complete physical freedom of access throughout all realms and stages of the cosmos. Images of the Sage possessing universal access to all realms are used to portray these far-off journeys, which seem like direct legacies of early Chinese depictions of the shaman and the shamanic journey. The early Daoist visions of the far-off journeys of the Sage portray the absolute freedom of physical movement throughout the cosmos and the world. For these journeys, the Sage depends absolutely on the presence of the pristine Dao, which he must embody in order to be set loose. The journey of the Sage has two levels of signification: the Sage travels from an external here to an external there, but these journeys can also be understood to lead from an internal here to an internal there by virtue of the early Daoist tendency to internalize the cosmos. The soteriological progress of the Sage is, in fact, an act of identifying with the Dao, the cosmos, and the world, and the field of this identification is squarely centered on the physical body of the Sage.

The most common set of images applied to portrayals of the journeys of the Sage almost invariably center on passes or gateways. Images of the gateway appear several times in the *Laozi* and represent, arguably, the richest source of the symbolic content of the work as a whole. In this light, one wonders if the later Daoist tradition that spins out a long tale of Laozi going to the

Hangu Pass as he leaves the "world of dust" is itself a redirected fictionalization of this image of the gateway.[57] The image commonly depicts the passageways between one realm and another in accordance with the different stages of the cosmological and soteriological sequences. In the context of the cosmological sequence, passage through a gateway leads from the pristine Dao to the ten thousand things; in the context of the soteriological sequence, passage through a gateway leads from the ten thousand things back to the pristine Dao. The Sage, by passing back and forth through these many gateways at will and at ease, enjoys complete access to all the realms designated with each stage. The first mention of a gateway occurs in *Laozi* 1:

> That-which-is-not names the beginning of Heaven and Earth, and that-which-is names the Mother of the ten thousand things. Thus, hold to that-which-is-not with the intent to witness its wonders; hold to that-which-is with the intent to witness its manifestations. These two (that-which-is-not and that-which-is) emerge together but are called by different names. Their sameness is called the mystery. Mystery upon mystery: this is the Gateway of All Mysteries.[58]

Laozi 6 also discusses another gateway, which in this passage is depicted as a vagina. The vaginal image is extremely apt because, like a gateway, it gives access in two directions: access from the outside moving in describes the potential potency of the sexual act and is associated with the soteriological potency of the Sage making the journey back to the pristine Dao, and access from the inside moving out describes the realized potency of the birthing process and is associated with the cosmological potency of the generation and emergence of the world. In this passage, the image of the valley also calls to mind the general anatomy of the female sex organ, associations completely in keeping with the general tendency of early Daoist discourse that consistently represent the pristine Dao as the Mother: "The valley spirit never dies, she is called the Mysterious Female. The Gateway of the Mysterious Female is called the root of Heaven and Earth."[59] Even though *Laozi* 1 and 6 describe these gateways by applying the same term "mystery" 玄 (*xuan*), these two gateways clearly describe two different gateways; the first is located in the space between the two modes of the pristine Dao as that-which-is-not and that-which-is, when the Dao existed alone in its own emptiness. This movement in turn represents the generation of *qi* as the breath or wind of the pristine Dao, and marks the birth of the One in the cosmological sequence of *Laozi* 42. The second gateway is located in the space between the Two and the Three, between the *yin-yang* and the world; it is the "Root of Heaven and Earth." Can we assume the presence of gateways standing at each of the transition points between the separate realms of the cosmological sequence?

A primary feature of these gateways is their sexual content that expresses the vital potency of the passage through them, whether this is the cosmological passage from the inside out or the soteriological passage from the outside in. The soteriological passage is differentiated from the cosmological passage not only by the direction taken through the gateways, but also by the fact that it is the Sage alone who can successfully make that passage. The possibility for the access of the Sage through these gateways is a common feature of early Daoist discourse, and this is also indirectly presented in the *Laozi* several times. A further common feature of these early Daoist depictions of the Sage passing through the gateways is that the Sage must himself become either an infant, or a female, or both an infant and a female; this is because the passage through a gateway is an event that is understood to be vitally potent for life in the same way that the sexual act and the birth process are vitally potent for life; the term *de* 德 is often used to denote this vital potency for life. These themes lie at the heart of *Laozi* 10: "In concentrating your *qi* and making it soft—can you make it like that of a child? . . . In opening and closing the Gateway of Heaven—can you play the part of the female? . . . Give birth to them and nourish them . . . this is called profound *de*."[60] This passage explicitly calls on the Sage's ability to become both an infant (to pass through the gateways) and a female (to provide birth and nurture to all things).

The Sage merges with the world, unites with and embodies the Dao, and enjoys universal access to all realms; this characterization is specifically Daoist and strictly differentiates the early Daoist Sage from the characterizations of the Sage belonging to all other early Chinese discourses, in which Sages rarely leave the ground. These several themes are present even in the earliest layers of early Daoist discourse. The *Neiye* 內業, often identified as one of the earliest writings of early Daoism, also structures its discussion of the Sage around this complex of themes; *Neiye* 13 states: "The Dao fills the entire world. . . . With one word of elucidation, above you reach up to Heaven; below you stretch down to Earth; and in between you pervade the nine regions."[61] The *Neiye* introduces the metaphor of the wellspring 泉原 (*quanyuan*) into its discussion of the Sage's journeys; here the wellspring signifies that the complete presence of the elements of the pristine Dao continually gush forth from within the physical body of the Sage, an image reminiscent of *Laozi* 4: "The Dao is empty, yet when you use it, you need never fill it up again."[62] The pristine Dao is inexhaustible, and its embodiment results in a total transformation of the body, allowing it to roam at will throughout the universe; thus states *Neiye* 8: "When the wellspring is not drained, you can freely circulate throughout the nine borders. You can then exhaust Heaven and Earth, and spread over the four seas."[63] The *Neiye*'s inclusion of this complex of themes centered on the Sage and his mergence with the world, embodiment of the Dao, and subsequent universal access, are concluded in the following passage: "If a human is aligned and tranquil . . . he will be able hold up the Great Circle (of Heaven) and firmly tread on the Great Square (of Earth)."[64]

The *Neiye* writings most likely predate the earliest writings of the *Laozi*; the *Laozi*, however, reinscribed the *Neiye*'s indications of the Sage into the complex vision of the world structured around the cosmological model set forth in *Laozi* 42. After the appearance of the *Laozi*, the writers of the

Zhuangzi and the *Huainanzi* greatly expanded the significatory field encompassed by the Sage; these writings most fully exploited the descriptive possibilities indicated in the minimalist rhetoric of the *Laozi*. Whereas the *Laozi* consistently designates the Sage as *shengren*, the *Zhuangzi*, *Huainanzi*, and other early Daoist writings employed a much richer technical vocabulary to explore the latent fields of his signification, as seen in the use of such terms as Spirit Person 神人 (*shenren*) and Genuine Person 真人 (*zhenren*). The word *shen* refers to the transformation of the physical body of the Sage, not to the transcendence from the physical body often associated with the English term "spirit;" the continued possession of the physical body is precisely what differentiates the Sage from a ghost.[65] The word *zhen* refers to the physical potentialities inherent in the body that the Sage realizes or perfects, not to the moral perfection of the will often associated, again, with the English term "perfection."

Zhuangzi 1 offers several characterizations of the complex of themes centered on the Sage, namely his merging with the world, embodiment of the Dao, and subsequent ability to journey freely throughout the universe. The first passage of this section describes the journeys of the important Daoist figure Liezi, who "journeyed with the winds for his chariot, clear, crisp and carefree, and he did not come back for fifteen days."[66] Even as good as Liezi is, he still depends on something in order to journey: the wind. The journey of Liezi is immediately juxtaposed with the journey of the Genuine Person: "As for the one who rides in a chariot of the transformations of the Six Qi, steering a true course between Heaven and Earth to travel into the infinite, is there anything that he depends on?"[67] This passage describes the utter freedom enjoyed by the Sage. Later in the same chapter, there is the *Zhuangzi*'s most celebrated characterization of the Sage, in which the theme of absolute freedom of movement is most notable:

> In the far-off mountains of Guyi, there lives a Spirit Person whose skin and flesh is like ice and snow, and who is gentle as a virgin. He does not eat the five grains but sucks in the wind and drinks the dew. He rides the *qi* of the clouds, yokes flying dragons to his chariot, and roams beyond the four seas. When the spirit in him is concentrated, it keeps creatures free from plagues and makes the grain ripen every year.[68]

The confluence of the themes relating the pristine Dao, the Sage who embodies this Dao by way of uniting with the world, and the images of gateways leading to all realms of the universe, are given two particularly striking presentations in the so-called outer chapters of the *Zhuangzi*. The first appears in *Zhuangzi* 23:

> There is something from which we are born, into which we die, from which we come forth, and through which we go in. That from which we come forth and into which we will go without manifest-

> ing its form is called the Gateway of Heaven. The Gateway of Heaven is that-which-is-not. The ten thousand things emerge from that-which-is-not. That-which-is cannot become that-which-is by way of that-which-is, it must come forth from that-which-is-not, and that-which-is-not always remains that-which-is-not. The Sage lodges himself therein.[69]

This passage is remarkable in many respects. The Gateway of Heaven depicts the passage from that-which-is to that-which-is-not in its presentation of the cosmological passage from the inside out, and it does this again by relying on the images of birth. Further, the Sage's ability to pass through this gateway is directly expressed, and this represents the soteriological passage from the outside in. *Zhuangzi* 12 gives another remarkable presentation of the gateway, relying again on the dialectic of that-which-is and that-which-is-not.

> In the Ultimate Beginning there was neither that-which-is nor names; there was only that-which-is-not. From this the One originated: there was the One but it was yet without form. That by means of which things are born is called the *de*. In that which was yet without shape, there was a breach that circulated without interval: this is called Destiny. Creatures were born from amidst respite and activity, and principle was born with the completion of beings: this is called Form. Form and body protect the spirit, each having its own normative standard: this is called Nature. A cultivated nature reverts to *de*. Extreme *de* is identical to the Beginnings. Identity is the Void, and the Void is the Great, like a bird's twitter. If [one's spontaneity] accords with a bird's twitter, it will accord with Heaven and Earth. Such an accord has no design, as if foolish and obscure: this is called Mysterious *de*, the same as the Great Flow.[70]

This passage begins with a partial recapitulation of the cosmological sequence of *Laozi* 42: the "Ultimate Beginning" refers to the pristine Dao that preceded everything, designated as "that-which-is-not." The One, designated in this passage as *de* instead of *qi*, is born from the depths of the pristine Dao. The One differentiates, and its division gives birth to "respite" 留 (*liu*) and "activity" 動 (*dong*) referring to the passive and active modes of the Dao as *yin-yang*. The Two differentiated and gave birth to Heaven-the Human-Earth; in this passage, this is referred to as the beings that are completed with Form—namely, the world. The "beings" categorized under "Form" signify the three realms of the world, and are not the same as the "beings" categorized under "Nature," which signifies the ten thousand things. The Three as the world gives birth to all phenomenal beings, by endowing them with a body formed from Earth and a spirit preserved from Heaven. Having body from Earth and spirit from Heaven, human beings in particular, because they represent the highest culmination of the cosmological sequence ending with the ten thousand things, are able to cultivate their nature, which is generated directly from the components and mechanisms of the cosmological sequence from the pristine Dao.

The cultivation of these cosmic components inherent in their nature allows a human to return to the One, the *de*, by way of the soteriological sequence. Restoring the One to the plentitude of its vital potency within the physical body (note the internalization of the sequence here) results in the spontaneous union with the pristine Dao itself. When union with the Dao has been accomplished, something very interesting comes to pass: a series of passages from the Void 虛 (*xu*) to the Great 大 (*da*). These passages represent the success won by the Sage through his successful completion of the soteriological sequence, and signify the attainment of universal access. The "Void" here refers to all cosmological stages preceding the generation of form, or that-which-is-not as "wonder" 妙 (*miao*) in the words of *Laozi* 1. The "Great" here refers to all of the realms and stages that include and postdate the generation of the world, or that-which-is as "manifestation" 徼 (*jiao*) in the words of, again, *Laozi* 1.

After affirming the universal access of the Sage throughout all realms of the cosmos and the world, the *Zhuangzi* passage immediately employs the image of the gateway, albeit in a very striking way: the beak of a bird. The opening and shutting of the bird's beak describes the gateways that stand between the different stages of the cosmology. The *Zhuangzi* appears to locate this passageway in the interval between the Void and the Great, and explicitly likens the opening and closing of the beak to the opening and shutting of Heaven and Earth. This should not be understood as referring to the gateway between Earth and Heaven, never discussed in the early Daoist writings, but to the gateway between the realms of the world (Heaven-the Human-Earth) and the realms of the cosmos from which the world came to be—in other words the realm of the pristine Dao before form. The metaphor of the bird's beak used to describe the gateway again calls upon the sexual symbolism of the vagina that gives a passage in two directions, and further can be associated with the symbol of the *vagina dentata*. This highlights the extreme danger of the passage, one that only the Sage can undertake without harm to the body.

These images appear in a related way in the final passages of the story of Huangdi and Guangchengzi from *Zhuangzi* 11. Guangchengzi, immediately after explaining his methods of cultivating the body in order to experience union with the pristine Dao, falls into a rhapsody in which he describes his far-off journeys to Huangdi.

> I will teach you about the Perfect Dao . . . For you I have ascended above the Great Luminaries, even to the source of the utmost *yang*. For you I passed through the Gateway of Profound Obscurity, even to the source of the utmost *yin*. . . . I preserve the One, residing in its harmony, and thus I have cultivated my body for 1,200 years. . . .

> Now I am about leave you to pass through the Gateway of the Inexhaustible and roam in Fields of the Unlimited. I will form a triad of luminosity with the Sun and the Moon; I will become constant with Heaven and Earth. When you are near me I am a blur, when you are far from me I am a blank! If all humans in the end will die, will I alone remain?[71]

This passage presents a vision of the Sage's ability freely to pass through all realms of the cosmos, and it touches on virtually all the major themes examined so far in this chapter. Guangchengzi unites with the pristine Dao through a reversion to the One. His union with the Dao endows him with extreme longevity: Guangchengzi has lived for 1,200 years. It also allows him to accomplish the second-order harmony that completes the world represented in the activation of the triadic relationship of Heaven-the Human-Earth; in the inspired words of Guangchengzi, this triad is represented by the Sun, pure *yang*, or Heaven; the Moon, pure *yin*, or Earth; and the Sage, equal parts *yin-yang*, or the Human. All these images find expression through the Sage's enjoyment of absolute freedom of physical movement represented by the image of the gateway, used twice in this passage. The first usage is particularly interesting for its associations with "the Gateway of the Mysterious Female" from *Laozi* 6. This gateway is explicitly associated with *yin* in the *Zhuangzi* passage; it is "the source of utmost *yin*." However, instead of the Mysterious Female (*xuan pin*) 玄牝 mentioned in the *Laozi*, Guangchengzi speaks of the Profound Obscurity (*yao ming*) 窈冥; still, what each phrase signifies is virtually identical: both call up associations with the female sex organ. The second usage of the metaphor of the Gateway is reminiscent of the first chapter of the *Zhuangzi*, where Zhuangzi explains to Hui Shi what he should do with a huge tree given to him. "Why not plant (the tree) in the District of Nothingwhatever, in the Fields of Extensive Obscurity, and go roaming away to do nothing at its side or ramble around and fall asleep in its shade?"[72]

The *Huainanzi* is replete with portrayals of far-off journeys undertaken by a myriad of Sages that share in this common complex of themes and images. The first chapter of the *Huainanzi*, *Yuan Dao* 原道, provides two striking descriptions of these journeys. In them, we witness the early Daoist penchant for calling upon the images of the shamanic journeys. In the first passage, the journeys of two kings, Tai Huang and Gu Huang, are depicted. These two kings are commonly identified with Fu Xi and Nu Gua, who represent the first humans.

> The two kings Tai and Gu got hold of the handle of the Dao and established themselves in the center. Their spirits roamed in transformation to bring peace to the four corners. Thus they were able to circulate like Heaven and be planted like Earth. Spinning on the wheel without cease, flowing with the water without stop, they ended and began again with the ten thousand beings.[73]

In the next passage, the two figures mentioned, Peng Yi and Da Bing, are commonly identified as water gods from early Chinese myth; here we see them appropriated in the early Daoist writings as ancient Sages. Commentaries to this passage tend to identify the Gates 排 (*pai*) of Changshe mentioned here with the *Zhuangzi*'s Gateway of Heaven discussed earlier. Several other images in this passage can be traced directly to similar descriptions found in the *Zhuangzi*.

> Long ago, in their charioteering Peng Yi and Da Bing rode the Thunder Chariot and harnessed the Six Cloud Dragons. They roamed about in the fine mists and galloped around in the hazy and nebulous. Ever more distant and ever higher, they made the Supreme Journey. They passed over frost and snow, leaving no tracks, and under the shining sun they cast no shadow. Swirling in typhoons they ascended the Ram's Horn spiral. Negotiating mountains and rivers, they leaped clear of Mount Kunlun. They pushed open the Gateway of Changshe and passed through the Gateway of Heaven.[74]

Early Daoist discourse is constructed on a Dao-based cosmology, and the reason that the Sage is able to journey far-off is that he unites with Heaven and Earth, and through successively passing through the gateways of the separate realms, comes to possess and embody the pristine Dao. Returning back to the phenomenal realms, and energized by the physical possession of the Dao, the Sage roams at will with absolute freedom of physical movement. Armed with the Dao-based cosmology, the destinations of the Sages lie not only in far-off lands within the boundaries of Heaven and Earth, but also in formless, placeless domains; their destinations are either within the spatial realms of Heaven-the Human-Earth, or the nonspatial realms preceding the world. Finally, the journeys of the Sage have concrete soteriological consequences both for his own physical body and for the world as a whole.

CHAPTER FOUR

EARLY DAOISM AND ONTOLOGY

THE FRACTURED WORLD

The world described in early Daoist discourse typically consists of the three realms of Heaven, Earth, and the Human: Heaven most generally refers to the celestial field, Earth refers to the geographic environment, and the Human refers to the social and cultural arena in the widest sense. Heaven is the sky, including the sun, moon, planets, and stars; Earth is the mountains and seas and all natural objects and creatures lying between the peaks of mountains and the bottom of the sea; and the Human is all that which defines and describes culture and society, including language, literature, all forms of knowledge, and the production of social, epistemological, and emotional distinctions. According to the cosmological theory implicit in this discourse, the three realms are in direct alignment at the generative moments of their coming-to-be, and they maintain this direct alignment throughout the continuing unfolding of the first-order harmony.

The earliest presentation of these ideas occur in *Laozi* 42, which depicts, in its minimalist fashion, the cosmological sequence beginning from the originary presence of the pristine Dao to the emergence of the three realms, culminating in the phenomenal reality of the ten thousand things. In the cryptic words of the *Laozi*: "Dao gave birth to the One, the One gave birth to the Two, the Two gave birth to the Three, and the Three gave birth to the ten thousand things."[1] *Laozi* 42 continues: "The ten thousand beings carry *yin* on their back and embrace *yang*, and through a blending of [these two] *qi* arrive at a state of harmony."[2] The last lines of this passage isolate any individual in general and describe, again in very cryptic terms, the soteriological sequence from the ontic here and now back to the full presence of the pristine Dao, resulting in the second-order harmony. In other words, two separate sequences are active in the processes of the completed world: a cosmological order of formation and a soteriological order of reversion. The ultimate union with the pristine Dao achieved through reversion is consistently described in terms of human physiology: the Sage acquires the Dao 得道 (*dedao*) or embodies the Dao 體道 (*tidao*).

The two sequences can be more fully described by saying that the birth and beginnings of the world (Heaven-the Human-Earth) culminate only in the first-order harmony, the partial and provisional fulfillment of the full potentiality of its own completion. The Sage embodying the Dao in the world necessarily realizes the completion of the world by reinstating a higher union of Heaven-the Human-Earth in the second-order harmony: he completes the being of the world. A less than perfect harmony stands between the first- and second-order harmonies, and this is a partial, fractured, and fragile harmony caused by the intentional activities of human beings. Intentional activity creates borders that separate Heaven-the Human-Earth, causing the initial harmony of the world to approach its breaking point. The Sage dissolves the borders that separate Heaven-the Human-Earth and opens the way for their mergence as one in the pristine Dao that is never fractured. The space separating the first-order harmony (cosmology) from the second-order harmony (soteriology) is filled by the present, fractured state of the world (ontology). Ontology, or the study of the present state of being of the world and the way things are within it, is a primary theme of early Daoist discourse, concerning as it does their ideas about the etiology of the breakdown of the first-order harmony.

Images of separation, divorce, and difference are prevalent in early Daoist discussions of the world in its ontologically fractured state. The terms "borders" 封 (*feng*) and "divisions" 分 (*fen*) are common in early Daoist writings, and they never show up in positive ways. As long as borders stand between Heaven-the Human-Earth, their completion will never be realized.

Transformation 化 (*hua*) signifies any ontological change in the state of anything that exists, and it also describes the mechanism whereby borders disperse and vanish, thus allowing direct access to the pristine Dao. Transformation is a naturally spontaneous event occurring within the world; examples of natural transformation abound in early Daoist writings because it was recognized as the Dao's most awesome power. However, the mechanisms of natural transformation are extremely delicate and easily disrupted by human interference. Disruptions are the consequence of intentional activity that diverges away from the initial harmony of the natural movements of the world, and it creates intervals, stoppages, and discontinuities in the spontaneous flow of change and existence between realms. These intervals create spaces in which are produced or inserted borders, divisions, and boundaries within the movements of the Dao, the circulation of *qi* in the body and the world; in this case, the simple English translation of *qi* as "air" or "atmosphere" is not entirely out of place. Borders standing between separations are maintained by continuing acts of human interference in the world. A portrayal of this occurs in *Zhuangzi* 26:

> In general, the Dao does not wish to obstruct [anyone]. For when a person is obstructed, he gasps; if he gasps, irregularities arise. When

> these irregularities arise, they bring with them a multitude of harm. Everything that is conscious depends on breath. However, if breath does not abundantly flow, this cannot be blamed on Heaven. Heaven seeks to make breath course throughout the day and the night without obstruction; human beings are to blame for the stoppage of natural circulation.[3]

Borders and divisions not only keep the three realms of the world separate, but also dissociate human beings, both individually and as a species, from the Dao. Borders and divisions originate in the imposition of distinctions in the world in which no distinctions exist naturally. The initial production of distinctions by human beings in the not-so-distant past marks a decisive turning point in the ontological state of the world, initiating a kind of ongoing fall in which the spontaneous movements of the Dao are disrupted then bordered, and in which human beings separate from the Dao more and more. Ultimately, of course, everything is the Dao, but in the fractured state of the ontological world, things that exist do so only as pieces or shards of the pristine Dao far removed from its full presence. The imposition of borders on phenomenal reality is essentially the imposition of borders on the Dao itself, insofar as phenomenal reality is one primary field of the being of the Dao. Human acts of imposition directly cause the breakdown of the first-order harmony of the world by disrupting the natural mechanics of spontaneous transformation, tearing that harmony apart. It is at this point that the presence and agency of the Sage is required in order once again to unite all aspects of the divided being of the pristine Dao in phenomenal reality, thereby allowing the Dao to resume its spontaneous movements of transformation and life-giving, life-sustaining potency. The Sage dissolves the borders that hinder natural transformation, and thereby sets the Dao free in the completed world of the second-order harmony.

SPLITTING BINARY DIFFERENCES: THE ONTOLOGICAL VISION OF THE *LAOZI*

The first-order harmony is spontaneous, natural, and extremely delicate, eminently liable to disruption by human interference. The second-order harmony is fundamentally therapeutic in nature, and its effectuation initiates a higher-level harmony than that achieved by natural processes alone. The Sage realizes the potential being of the completed world, and his presence is necessary for it because of the damage caused by human intentional activity. Intentional activity objectifies the realm of the Human and brings it to a state of hypermaturity that threatens to separate it from Heaven and Earth once and for all. The substantial cause for this is the production of distinctions and borders, the negative effects of human intentional activity, which dramatically increases the possibility that the Dao will vanish from the world; conversely, it also dramatically increases the scope of the futural harmony that can now be effectuated by the Sage, because this harmony must now include human beings as a necessary part of the world. The inclusion of human beings in this project of completing the world is what gives the tremendous power to the soteriological sequence.

In early Daoist writings, human beings are seen as supremely creative and productive. Of the three realms that make up the world, the realm of the Human is the only one that, originally forming through the spontaneous unfolding of the cosmology, is brought to maturity through human production and human labor. The realm of the Human is synonymous with the arena of culture and society, and is open to human development. Once this development reaches such an extreme that it impedes the free movements of the Dao, the harmony of the world progressively deteriorates and the three realms close off from each other. Ancient culture heroes, such as Huangdi, Yao, and Shun, first introduced the techniques and institutions that made human culture possible, and this alone affected the free movements of the Dao, though only to a slight degree. As cultural techniques and institutions attain a developed degree of autonomy from the realms of nature, so culture increasingly dissociates from the natural rhythms of the pristine Dao. The realm of the Human starts to exist independently from the realms of Heaven and Earth; absolute independence from nature would mean the death of both nature and culture, and this is the main threat that intentional activity poses to the world. What I call the ontology of early Daoist discourse is structured around the recognition of this threat to the harmony of the world. The separation of the realm of the Human from the realms of Heaven and Earth is only one aspect of this threat; more fundamentally, what is at stake is the separation of human existence from the life-giving, life-sustaining presence of the Dao. Early Daoist writings give a great deal of attention to making transparent this separation between the realm of the Human and the Dao, and locate the cause for it in the production and insertion of borders.

Whereas the notion of borders is made explicit throughout the *Zhuangzi*, the *Laozi* brings them up in its own characteristically minimalist way. The *Laozi* often phrases its discussion of the growing separation between human beings and the Dao in terms of originary binary distinctions arising from within the mental, emotional, and sensory fields of the human constitution. Acts of simple originary distinctions, such as high/low and hot/cold, gradually disseminate into the social spheres of human activity, where they usually translate into distinctions of social hierarchy and proper virtue. Simultaneously, with the dissemination of these distinctions to cover the whole gamut of human experience on the social sphere, human beings construct and produce the properly human world of culture and society. If this were to progress unchecked, it would result in the total separation of the autonomous realm of the Human.

Passages describing the separation of human beings from the pristine Dao abound in the *Laozi*. *Laozi* 23 clearly attributes this separation and loss of unity with the Dao to human activity:

> Therefore, one who takes up the project with the Dao is one with the Dao. One who is committed to *de* is one with *de*. One who is committed to displacement is one with that displacement. One who is committed to the Dao, the Dao also happily receives; one who is committed to the *de*, the *de* also happily receives; one who is committed to displacement, that displacement also happily receives.[4]

Culpability for the present, deficient state of the world is attributed to human beings pursuing activities in ways that displace their initial harmony with the natural course. The cause of this is to be found in the distinctions created by human beings, which originate in reflections on the experience of the natural world within the sphere of internal consciousness. On the level of instinct, the simplest distinction is between agreement (like) and rejection (dislike); on the sensory level, it is that between beauty (attraction) and ugliness (repulsion); and on the level of consciousness, the most critical level, the simplest distinctions are between self and other, and life and death. According to *Laozi*, all of these are not only false distinctions, but they are also the root cause for the disharmony of the world. Imposed on the experience of reality, these distinctions have radical consequences: they border the world and impede the Dao. Left unchecked, distinctions breed further distinctions with no end in sight. *Laozi* 20 says, "Agreement and rejection—how great is the difference between them? Beauty and ugliness—what is it like, their difference? The one who is feared by others must because of this fear other men. Wild, unrestrained! It will never come to an end!"[5] The *Laozi* throughout assumes the original unity of all things in the world within the Dao, and this unity is disrupted by distinctions arising from discriminations. These discriminations are, in turn, the consequence of differences arising from thought (agreement and rejection) and sensory experience (beauty and ugliness).

This theme also occurs in *Laozi* 2, where it argues that all distinctions originate as binary pairs created from the flux of ontological reality: "Therefore, that-which-is and that-which-is-not are mutually produced; difficult and simple are mutually completed; long and short are mutually formed; high and low are mutually filled out; tone and voice are mutually harmonized; front and back mutually follow each other constantly."[6] The dissemination of these distinctions displaces the first-order harmony of the unitary world, and it is up to the Sage to unite with the Dao and achieve the second-order harmony, rendering the world a unitary whole once again. *Laozi* 12 presents a curt description of the activities of discrimination run amok, and the passage ends on a positive note directed to an aspect of the project of the Sage in relation to a second-order harmony.

> The five colors cause the eyes to go blind. The five tones cause one's ears to go deaf. The five flavors confuse one's palate. Racing horses and hunting cause one's mind to go mad. Goods that are hard to obtain pose an obstacle to one's travels. Therefore in the order of the

Sage—he is for the stomach and not the eyes; he rejects that and chooses this.[7]

This passage has two parts. The first part describes the displacement of the genuine center of a person, which ideally ought to be kept within. The "fives" that it mentions refer to the dazzling variety of objects that cause sensory gluttony leading to overload in relation to sight, taste, and sound; these, together with the excitement of racing and hunting and the desire to obtain possessions, describe a highly refined taste developed by polite education. Pursuing these tastes causes a person to attend to external goods and matters instead of to the physical body. The second part of this passage indirectly describes two kinds or fields of discriminations, and the method of overcoming them. The first are discriminations of experience, overcome by prioritizing the stomach, the physical center of one's person, over the eyes, the most insidious passage through which one displaces the physical center in pursuit of external objects and satisfactions. The second are the discriminations of thought, overcome by prioritizing "the rejection of that" 去彼 (*qubi*), referring to external objects and satisfactions, and "the choosing of this" 取此 (*quci*), referring to what is right here in the physical center. In other words, the Sage rejects everything that is "that" or "other" in relation to "this" or "I," and chooses "this" insofar as the "this" refers to anything that is present here at hand in the moment and at the center. Thus, the project of the Sage is precisely to restore the unity of the Dao by a direct overcoming of the discriminations of thought and experience as they are presented in *Laozi* 2 and *Laozi* 20.

Whereas the Sage, in overcoming discrimination at the root, "rejects that and chooses this," the opposite is generally true for ordinary human beings. To choose "that" is to reify distinctions, often no more than the pure products of human thought, within the phenomenal world of human activity, thereby creating differences that are maintained by borders. The first two lines of *Laozi* 2 describe these sorts of reifications, and depict their spread within the field of unitary being: "When everyone in the world knows the beautiful as beautiful, ugliness comes into being. When everyone knows the good, then the not good comes into being."[8] Constant acts of discrimination create empty gaps in the free flow of existence, and the gap opened up by the difference between "beautiful" and "ugly" is simply one example among an infinite number of gap-creating discriminations. These gaps are held in place by the insertion of borders between discriminated binary pairs, and the empty space opened up by the difference between the binary pair "beautiful–ugly" is, again, one example of a border placement. Such borders sharply curtail the free movement of the Dao, and impede natural transformation.

Acts of discrimination, unique to human beings, are products of intentional activity, what the *Laozi* calls *wei* 為 or *youwei* 有為. This is negatively contrasted to non-intentional activity, what the *Laozi* calls *wu-wei* 無為, which is devoid of discrimination, expectation, and goal-orientation. Intentional activity is human behavior that actively participates in the production of constantly expanding artificial sectors of experience that gradually impinge on the natural realm and, when left unchecked, threaten to dominate it such that nothing of that natural realm can remain in a natural state. The artificial sectors of experience altogether make up the realm of the Human, which now is no longer in a harmonious relationship with the realms of Heaven and Earth. Early Daoist writings give many descriptions of the differences between intentionality (*youwei*) and non-intentionality, (*wuwei*); one example that makes this particularly clear is from *Huainanzi* 1.

> The Sage . . . quiescently acts non-intentionally (*wuwei*), yet nothing is left undone (*wubuwei*). He serenely does not impose order on anything (*wuzhi*), yet nothing is left unordered (*wubuzhi*). By "acting non-intentionally" is meant not being ahead of things in taking action; by "nothing left undone" is meant making use of what is done by other things; by "not imposing order" is meant not putting in a substitute for what is spontaneously so; by "nothing is left unordered" is meant making use of the mutual causation that obtains among things.[9]

Laozi 48 briefly contrasts intentional and non-intentional activity. Intentional activity, for the *Laozi*, is what cultured humanity specifically intends to have its members develop through the study and cultivation of artificial behavior. *Laozi* 48 says:

> Those who pursue study daily increase; those who pursue the Dao daily decrease. They decrease and decrease until they reach a point where they act non-intentionally. They act non-intentionally and nothing is left undone. Somebody who takes control of the world must constantly be non-involved in worldly activities. If he becomes involved, he becomes unworthy of taking control of the world.[10]

The *Laozi* consistently rejects all cultural programs of social, political, and ethical cultivation, and it denigrates the core collection of "intentional" virtues that have come to be identified with the Confucian tradition. The breakdown of the harmony of the world, at the expense the Dao and the *de*, is the cost of such programs. *Laozi* 18 states, "Therefore, when the great Dao is rejected, then there is humaneness and righteousness. When knowledge and wisdom disappear, then there is great hypocrisy. When the six relations are not in harmony, then there is filial piety and compassion. When the country is in disruption and chaos, then there are virtuous officials."[11] The humaneness, righteousness, and filial piety in this passage are three of the supreme virtues extolled by the Confucian tradition and their cultivation is a prerequisite for participation in Confucian society. According to the *Laozi*, the proper exercise of these virtues can only be carried out by an endless series of exacting discriminations, and *Laozi* 25 forcefully claims that these intentional virtues have no part in the natural order of the world: "Heaven and Earth are not humane. . . . The Sage as well is not humane."[12] Here, "humane" 仁 (*ren*) describes the intentional virtue espoused by the Confucian tradition that represents their conception of the highest moral good, but early Daoist writings commonly reject the Confucian claim for the naturalness of this virtue, identifying it instead as an artificial distinction of human thought and behavior.

In the vision of the *Laozi*, the gaps in what was once the unified fabric of the world are caused by human acts of distinction. These gaps are held open by the intentional virtues that are inserted within them, thus constituting the most insidious bordering principle that widens the separation between the realm of the Human and the realms of Heaven and Earth. *Laozi* 19 describes the positive results of erasing the bordering principle:

> Eliminate sagehood and throw away knowledge, then the people will benefit one hundred times over. Eliminate humaneness and throw away righteousness, then people will return to being filial and compassionate. Eliminate artistry and throw away profit, then there will be no robbers and thieves. These three sayings are not complete as a text, and I append the following: Manifest plainness and embrace the genuine; lessen self-interest and reduce desires; eliminate learning and be without anxieties.[13]

Many of the intentional virtues singled out by the *Laozi* in this passage are extolled as natural in others; for example, the *Laozi* sometimes praises sagehood 聖 (*sheng*), but at other times denigrates it. It is important to note that the different judgments of these virtues are either positive or negative based on their use within a specific context in a specific passage. Having said that, the six intentional virtues given in the received version of *Laozi* 19, namely sagehood (*sheng*), knowledge 智 (*zhi*), humaneness 仁 (*ren*), righteousness 義 (*yi*), artistry 巧 (*qiao*), and profit 利 (*li*), are interesting to compare with the six intentional virtues singled out in the corresponding section of the Guodian *Laozi*.[14] That is, the Guodian *Laozi* lists knowledge (*zhi*) and artistry (*qiao*) as that which produces "goods hard to obtain," and includes profit (*li*). But instead of humaneness, righteousness, and sageliness, it lists distinctions 辨 (*bian*), deliberation 慮 (*lu*), and transformation 化 (*hua*). In this instance, "transformation" is used in the Confucian sense of moral transformation through education, and this stands in stark contrast to the early Daoist understanding of the same term, which typically refers to an ontological change of state of any phenomenal thing. The explicit mention of "distinctions" in the Guodian *Laozi*, later omitted and replaced with "humaneness" and "righteousness" in the Mawangdui and received editions, indicates a deepening awareness of the insidious activities of the intentional virtues extolled by early Confucian discourse, whose influence at the time of the composition of the Guodian *Laozi* had not been so recognizable. The most remarkable description of the genealogy of these intentional virtues extolled by early Confucians discourse appears in *Laozi* 38. This passage, however, attempts something more than simply uncovering the social creation of the Confucian virtues: it attacks the very act of discrimination itself.

> Highest *de* is not "*de*," and therefore is truly *de*.
> Lowest *de* never loses sight of "*de*," therefore is not *de*.
> Highest *de* is non-intentional, and is done non-intentionally.
> Lowest *de* is non-intentional, but is done intentionally.
> Highest humaneness (*ren*) is intentional, but is done non-intentionally.
> Highest righteousness (*yi*) is intentional, and is done intentionally.
> Highest ritual is intentional, and when someone does not conform to it, others then angrily roll up their sleeves and force them.
> Therefore, when the Dao is displaced, then there is *de*.
> When *de* is displaced, then there is humaneness.
> When humaneness is displaced, then there is righteousness.
> When righteousness is displaced, then there is ritual.
> As for ritual, it is the thin edge of loyalty and sincerity, and the beginning of disorder.
> Foresight is the flower of the Dao, and the beginning of ignorance.
> Therefore the Great Human dwells in the thick but not in the thin; dwells in the fruit but not in the flower.
> Therefore, he rejects "that" and chooses "this."[15]

In this passage, the *Laozi* again directly examines the gradual closing off of the Dao from the human world. As intentional acts are performed in ever more depraved ways, so does the world become ever more discriminated and bordered, cutting off the free movement of the Dao. There are several discrete sections to this passage. The first two lines draw a clear difference between the natural embodiment of a virtue, in which the agent is oblivious with regard to whether or not his acts fulfill the formal criterion that defines virtue as such, and the intentional performance of acts that are done with full awareness that they indeed do fulfill the formal criterion that defines virtue. Thus, the pure embodiment of *de* occurs when the agent possesses no self-conscious intention to have his acts fulfill any criterion of virtue: all of his acts are motivated by the spontaneous unity with the Dao and the *de*, with no self-conscious reflection stepping in to break that unity; this is "highest *de*." Self-consciously performed acts that the agent specifically intends to fulfill the formal criterion are not spontaneous expressions, but demonstrate a false and empty embodiment of that virtue; this is "lowest *de*." "Highest *de*" is non-intentional, and "lowest *de*" is intentional.

Emphasizing the relation of the Dao to non-intentional activity, and intentional virtues to intentional activity, this passage presents its genealogy of morals in the next five lines. Thus, the highest expression of *de* is spontaneous (non-intentional), and is performed without any intended goal. Lowest *de* is also spontaneously expressed, but this is performed with an intended goal—namely, the fulfillment of the formal criterion that defines it. Highest humaneness is performed intentionally and with conscious awareness; it is no longer a spontaneous expression, yet it is performed for its own sake, again with no specifiable goal in mind. Humaneness is the supreme virtue in Confucian discourse, and the *Laozi* does not completely condemn it, but sees it as already one step removed from the natural embodiment of *de*. Righteousness is one step removed from the expression of humaneness: it is performed with self-conscious intention, and with specific ends in sight. Having specific ends in sight means that this virtue is not performed non-intentionally, but in order that other people can bear witness to one's correct performance of it. As a virtue, the formal criteria going into the definition of correct performances of righteousness still have a standard that is at least partially internal, because it takes into account the mental state from which an action is performed. With the step to ritual, the standards of fulfillment are completely external, because this refers to specific acts judged solely on their conformity with public codes. Any given act is correct or not based on its physical execution alone. When people do not conform to the strict codes of ritual behavior, ranging from the manner in which one greets other people to the hand one uses in handing them an object, then social cohesion is destroyed and disorder emerges, and the only measure left to keep human society in line is force of arms.

The next several lines depict the gradual erosion of the spontaneous embodiment of natural virtue, expressed as the total absence of intention, which ultimately leads to the final solution of execution and warfare. This passage is brought to a close with a brief discussion of the Great Human 大丈夫 (*da zhangfu*), another term for the Sage: he dwells in the thickness of undivided, nonbordered being, allowing the undivided being of the Dao to enjoy once again freedom of movement throughout the world. The Sage rejects intentional acts and their associated artificiality, and nurtures the full being of what is directly present at hand in the moment. He performs *wuwei*, non-intentional activity.

Two further passages from the *Laozi* help to complete this picture of the role of the intentional faculty in relation to the setting apart of the human world as an autonomous realm detached from the Dao. *Laozi* 29 says,

> I see that one who desires to take the world and intentionally act on it will never obtain it. The world is a sacred vessel; it cannot be intentionally acted on, and it cannot be held. One who intentionally acts on it destroys it; one who holds it loses it . . . Therefore the Sage rejects the extreme, rejects the excessive, and rejects the extravagant.[16]

Laozi 64 can be read as a direct continuation of *Laozi* 29, but it offers additional insight into the destructive impact caused by intentional action to the harmony of the world; once again, the Sage is singled out as the single figure who can undo the damage: "Those who intentionally act on it ruin it; those who hold it lose it. Therefore the Sage acts non-intentionally, and as a result he does not ruin it. He does not hold it, and as a result he does not lose it. In people's handling of affairs, they always ruin it when they are right at the point of completion."[17] This passage draws together multiple themes, including the relation of intentional and non-intentional activity with the world: intentional activity destroys the fullness of the harmony of the world, while non-intentional activity alone is capable of completing the world. It is only the Sage who is capable of non-intentional activity and, thus, of completing the world. Ordinary human beings, separated from the Dao and thus from the fullness of the being of the world, have no possibility of completing the world but rather do just the opposite: their intentional discriminations act only to further border beings one from the other, resulting in further fragmentations of its unity.

The *Laozi* includes a series of descriptions portraying the brutality of the human world and consistently relates this to the loss of the Dao due to human acts of discrimination. *Laozi* 58 says: "When the government is muddled and confused, the people are genuine and sincere. When the government is discriminate and clear, the people are crafty and cunning."[18] Indeed, the *Laozi* takes the brutality of the human world, typically portrayed with biting anarchic insight, as one of the most important and devastating consequences of discrimination.[19]

HUMAN LABOR GETS A TURN: THE ONTOLOGICAL VISION OF THE *QIWULUN*

Zhuangzi 2, *Qiwulun* 齊物論, provides a detailed discussion centered on the bordered world, the origins of borders, and the ways to overcome borders and restore the lost harmony of the Dao with the world. Even the chapter title advertises this as its primary topic of discussion, regardless of how one chooses to translate it. A. C. Graham translates it by "The Sorting which Even Things Out"; Victor Mair by "On the Equality of Things"; Burton Watson by "Discussion on Making All Things Equal." The Chinese terms are *qi* (equalization, parity), *wu* (beings, things), and *lun* (explanation, essay, theory). The tenor of the title changes depending on whether one emphasizes *qi* or *lun*; if one gives emphasis to *lun*, then the title translates as "A Discussion on the Equality of Beings" (*qiwu:lun*); if one emphasizes *qi*, the translation is "Equalization of Things and Theories" (*qi:wulun*). Grammatically, nothing necessitates the reading of *qiwu:lun* to the exclusion of *qi:wulun*. Certain modern studies of the *Qiwulun* interpret this chapter as a sophisticated theory of relativism, in which all judgments and viewpoints are equal or equally relative.[20] My reading radically differs, beginning even with how I understand the meaning of the title that I take to refer to the early Daoist project of harmonizing (*qi* as parity in the sense of harmony) the unitary world (inclusive of *wu*, things, and *lun*, theories) that has become sundered and divided. Thus, I provisionally translate the title as "Harmonizing (*qi*) Objects of Experience (*wu*) and Theories About Them (*lun*)." Although my translation is primarily suggestive, it is not grammatically beyond board.

The *Laozi* passages discussed in the preceding section of this chapter reveal a deep concern about the presence of distinctions that impede and distort the movements of the Dao in the world. The terms of the *Qiwulun* resolve these distinctions into two categories: namely, mental distinctions of thought (*lun*) and experiential distinctions of phenomena (*wu*), though I strongly hesitate to impute any rigid internal/external dialectic to early Daoist discourse. The writer of the *Qiwulun* appears well aware of these ideas as presented in the *Laozi*, and thus I argue that the vision of the world on which the *Qiwulun* is anchored substantially squares with that of the *Laozi*. In relation to experiential distinctions of phenomena (*wu*), the separation of the being of the world through the imposition of borders is the root cause that gives rise to the perceived presence of individual beings as isolated entities. In relation to mental distinctions of thought (*lun*), individual beings perceived as isolated entities are the root cause that produce separations and borders in the movements of the Dao through judgment, discrimination, and language. The *Qiwulun* thus can be read as an extension of the writer's own ideas about the harmonization of the *Laozi*'s "this" and "that" that will, through the agency of the Sage, overcome the breaches and borders standing within the tattered remains of the once unified world.

The *Qiwulun* differs slightly in its understanding of the essential etiology of the bordering of the Dao. Whereas the *Laozi* attends more to the experiential nature of discriminations, "rejecting that, choosing this," the focus of the *Qiwulun* is more concentrated on diagnosing the issue from a perspectival standpoint based on positional reifications of self and other. In other words, both the *Laozi* and the *Qiwulun* assert that all distinctions and discriminations are ultimately false. The *Laozi* speaks of a choice between "this" and "that," and it advises people to step outside of the behavioral positions from which choices are produced, identified with intentional action (*youwei*), in order to act spontaneously in accordance with the standards of non-intentional action (*wuwei*). On the other hand, the *Qiwulun* prefers to attack the origin of the choice at the very root of individual positionality, thereby collapsing all distinctions in on themselves. This is nowhere made more assertively than in the phrase, "Without the 'that,' there is no I. Without the I, there is nothing by which to choose."[21] Here the *Qiwulun* not only makes the strong claim that all choices arise from a fundamental distinction between self and other, but that even what we take as the self is a false distinction arising from its opposition with any "that." The *Qiwulun*, however, does not mean to say that the individual is devoid of fundamental existence, but rather that the notion of the self, by way of self-identification, necessarily gives rise to the production of the distinction between self and any possible other that allows a person to construct an "I."

The opening section of the chapter draws a sharp distinction between the constructed self 我 (*wo*), and the authentic self 吾 (*wu*). In the opening passage, we read that Nanguo Ziqi 南郭子綦 (South Side Master Basis) reclined on his armrest while breathing deeply in some form of meditative practice, and he "let go of himself as if he had lost his counterpart 耦 (*ou*)."[22] Upon leaving his experience and returning to himself, Nanguo Ziqi says, "Just now I (*wu* as authentic self) lost my I (*wo* as constructed self)."[23] The import of this distinction is only partially semantic.[24] The *wo* should by no means be understood as referring to an objectifiable self, but rather as that sense of self that humans construct based on the distinction between "this" and "that" or "self" and "other"; *wo* is always relational with regard to what is not *wo*. Although the etymology of the term is unclear, it seems to represent a hand holding a spear, an image of conflict that is right in line with the *Qiwulun*'s understanding of the conflictual space between constructed self 我 (*wo*) and phenomenal other 彼 (*bi*). The text also discusses a second conflictual space between the authentic self 吾 (*wu*) and a spiritual other 耦 (*ou*), representing what seems a psychic "counterpart" within consciousness. For the *Qiwulun*, even psychic positionality is conflictual, and the opposition between *wu* and *wo* marks the originary distinction that gives rise to a notion of the self from which all other distinctions are deployed.

The following passage pursues the notion of the absence of any self-same psychic core that would authorize some enduring identity between the constructed self and the authentic self. It presents the ways in which a person finds himself present to himself, namely in transient and antithetical moods.

> Now delighted, now angry; now lamenting, now glad; now anxious, now sighing; now fickle, now obstinate; now modest, now willful; now insolent, now fawning: these are like music coming out of a void, steam forming into mushrooms. Day and night these mutually alternate with what came before and nobody knows from what they sprout.[25]

Moods as identity markers allow a person to identify and define himself to himself at any given moment, but they are not decisions or even realizations resulting from acts of intentional choice or pure will; rather, moods are just spontaneously present in accordance with any number of mental and environmental factors.[26] The *Qiwulun* next questions even the existence of any kind of psychic core or center of identity, which it hypothetically calls the "True Commander" 真宰 (*zhenzai*). The word I translate as "moods" is *qing* 情; although I am not familiar with other works that translate *qing* by "moods," in this instance the *Qiwulun* is describing precisely that.

> There appears to be a True Commander, but we can find no sign of it. We can follow its ways in true faith but we never see its body; it has moods but no body. One hundred joints, nine openings, six organs, all are complete and present within me; which should I

> recognize as more kin to me than another? Are you pleased with them all? Do you have a favorite among them? If so, then does it have the rest as its vassals and concubines? Are its vassals and concubines inadequate to govern among themselves? Do they take turns acting as each other's lord and vassals? Is there one among them who is the genuine lord? If you search for its mood and do not find it, this neither adds nor subtracts from its genuineness.[27]

This passage directly confronts notions of the self that were current in early Chinese philosophical traditions. According to the early Confucian discourse, the heart 心 (*xin*) was understood to be the ethical, deliberating, rational, and emotional core of the individual moral agent, what the *Qiwulun* calls the "True Commander." It not only says that if you search for this it cannot be located, but that even if there is such a True Commander, it never remains identical to itself from moment to moment. The *Qiwulun* here attacks the conditions for the constructed self referred to by the term *wo*; it never puts into question the ontological, brute presence of the person, referred to by the term *wu*. The *Qiwulun* is not formulating an argument of the relativity or skeptical nature of knowledge. It is not positing the presence of a higher self or a true self, but arguing more radically that any posited notion of self is a false distinction. There is only presence, and this presence, like the Dao, can only attain complete liberation in open space and time with the extinction of discriminations and the dissolution of borders.

The *Qiwulun* continues its analysis of the conditional nature of the constructed self by criticizing a conception of the completed body, defined by its ability to conform to ethical norms, that early Confucian discourse espouses. The completed body seems to be, in the view of the *Qiwulun*, a necessary adjunct of the constructed self. According to Confucian teachings, only when one has mastered the proper distinctions that undergird social virtues can one be considered a completed person with a completed body. By contrast, the *Qiwulun* laments the uncontrollable nature of things, which fail to be amenable to containment within the distinctions and borders of the judging mind as it attempts to dominate behavior. The *Qiwulun*'s image of the galloping horse is deeply reminiscent of *Laozi* 12 cited earlier, where it also describes the uncontrollable nature of things in the face of human efforts to dominate them. Noteworthy in the *Qiwulun* passage here are the images of being torn apart, used to portray the shredding of being.

> Is it not lamentable, the way the body comes up against things, how they cut each other apart, how they break each other down, the taking of its course ends up as if in a gallop which nobody is able to bring to a stop? Is it not grievous, the way the many bodies labor, labor, and do not see the completion of their works, exhaustedly, worn down by labors and not knowing that to which they will return? People call this state "not dead," but who gains from this? Its

> form changes, as does the heart along with it, and can this not be called the Great Sorrow? Human existence, is it not at bottom like this, a depletion? Or am I myself depleted, while others, however, exist who are not depleted? If people follow their completed heart and treat it as authoritative, then who alone is without this authority? Why must there be these [authorities] for those who possess the knowledge of transformation and allow their minds to choose of themselves? Idiots also possess this [authority]. For there to be the distinction between right and wrong before the completion of the heart is the same as "Going to Yue today and having arrived yesterday," as taking "that-which-is-not" as "that-which-is."[28]

For the *Qiwulun* as for the *Laozi*, the disruptions caused to the Dao are closely related to the problem of language. As *Laozi* 1 opens with a meditation on the Dao and names ("Dao's that are spoken are not constant Dao's, and names that are named are not constant names"), so *Qiwulun* asks, "What darkens the Dao so that there is the distinction between genuine and artificial? What darkens speech so that there is the distinction between assent and rejection?"[29] When both the Dao and speech are darkened, clarity is lost and with this come four central discriminations that consequently shatter to an even further degree the threatened harmony of the world: genuine 真 (*zhen*) and artificial 偽 (*wei*), and assent 是 (*shi*) and rejection 非 (*fei*). The Dao darkens, and this allows the production of the distinctions that bifurcate knowledge of the Dao into genuine and artificial. Names, or speech signification, darken, and this allows the production of the distinctions that tear signification apart in terms of assent and rejection. Calling upon the language of completion so prevalent in early Daoist discourse, the *Qiwulun* answers the posed question.

> The Dao is darkened by small completions and speech is darkened by illustrious and flowery [language], and it is from this that we have the Confucian and Mohist distinctions between assent and rejection. What one affirms the other rejects; what one rejects the other affirms. But if we want to affirm their rejections and reject their affirmations, then the best thing to use is illumination.[30]

My hypothesis is that "small completions" 小成 (*xiao cheng*) refer to all the discriminations performed by the self, including the originary discrimination of the constructed self over against everything other, together with all further discriminations of thought and language performed by it. Small completions in this instance characterize the assents and rejections by which is carried out Confucian and Mohist argumentation, proceeding step by discriminated step. Employing images of light and dark, this passage explains that the way to overcome the darkening of the Dao is simply to take a position outside of the constructed self and be immersed in the liberating illumination of the unified and unifying Dao. But the answer to the problem is a bit too simple, passing over as it does the incorrigible assertiveness of the constructed self with its intentional behavior. Thus, the *Qiwulun* immediately undertakes a further assault on the intentional, discriminative core, empty as it ultimately is (as the *Qiwulun* earlier said, "Human existence, is it not at bottom . . . a depletion"), which continues to establish the constructed self in confrontation with things to be discriminated. The following passage demonstrates the *Qiwulun*'s willingness to participate in philosophical argument that relies upon language distinctions in order to give an airtight logical formula that explains the origin of discrimination, called "the theory of mutual origin" 方生之說 (*fang sheng zhi shuo*).

> No thing is not other, no thing is not self.[31] Taking the position of the other you do not see yourself as the other, but when you take the position of the self then you know it [is other]. Therefore I say, the other arises from the self, and the self also follows from the other. This is the theory of the mutual origin of other and self.[32]

And yet to formulate such argumentative theories by relying on language distinctions does not enable one to restore the unity that preexists these language distinctions, for once such distinctions exist, there is no way to control them by means of language from which such distinctions were produced in the first place. Thus, immediately after formulating "the theory of mutual origin," the *Qiwulun* describes the futility of using language discriminations in the effort to restore the original unity of the Dao. It does this with an outrageous demonstration of the collapse of signification that has always already begun even at the very moment that one starts to use language distinctions: "Mutual origin and mutual destruction, mutual destruction and mutual origin, mutual acceptance and mutual negation, mutual negation and mutual acceptance. Following on yes and following on no, following on no and following on yes."[33]

Language discriminations chop up and border original, unitary reality, but language signification can only go so far before it runs up against the constant flux of changing reality, rendering signification useless after a certain point. The only way out of the quandaries of the breakdown of sense and signification that fragment the world lie in the purview of the Sage who embodies the pristine Dao in the full clarity of its unfragmented unity. Instead of pursuing ever deeper, ever more concise distinctions from the constructed self, the Sage allows all alternatives to proceed on their own, without discrimination.

> Because of this [the nonsensical consequences of argumentation], the Sage does not follow this way, but illuminates it from Heaven; this also follows from self. Now it is clear that the self is also other, and the other is also self. "That" also is one with assent and rejection, "this" also is one with assent and rejection. Is there now really an other and a self? Is there now really no other and no self? Neither

> other nor self obtains its counterpart: this is the pivot of the Dao. The pivot begins to obtain its middle point and through it responds to either without limit. Assent is also without limit, and rejection is also without limit. Therefore nothing is better than to rely on illumination.[34]

In the clarity of the Dao not darkened or bordered by distinctions, all things are harmonized; all things are one with the movement of the Dao, and it is from this movement unimpeded by the dislocations arising from distinctions that the clarity of the Dao shines forth in its full plenitude. The distinctions inserted into this movement of the Dao, reinforced through assent and rejection, construct the borders that impede the movements of the Dao in the world. Allowing 可 (*ke*) assent and rejection or not allowing 不可 (*buke*) assent and rejection to any choice created and posed by distinctions is the insidious mechanism by which discriminations and borders are imposed on reality. This plays out either in linguistic form in the intentional intellect (e.g., through allowing or not allowing a certain theory to be assented to or rejected), or in social form in the public arena (e.g., through allowing or not allowing a certain act to be assented to or rejected). All these distinctions manhandle the flux of reality through imposition and interference. The integral oneness of the Dao lies in stark contrast to the dividing intellect that shreds the unity of reality. These are the ideas expressed in the following passage.

> Allowable (*ke*)?-allowable. Unallowable (*buke*)?-unallowable. The Dao takes form through its movements; things are so through their being named. Why are they so? In being so. Why are they not so? By not being so. Things have that by which they are inherently so; things have that by which they are inherently allowable. There is no thing that is not so, and there is no thing that is not allowable. Therefore, even when acts of assent discriminate a stalk from a pillar, or a leper from Xi Shi [a paragon of beauty], things grotesque or strange, the Dao still penetrates them, making them one. Their ordering is their formation, their formation is their destruction. Yet all things, being inherently without this formation and destruction, again are penetrated and made one. Only one who has fathomed this realizes the penetration that makes beings one, and does not use intentional assent but lodges them in the ordinary. . . . For this reason, the Sage harmonizes them with both assent and rejection and rests on the potter's wheel of Heaven. This is called the "Double Walk."[35]

This passage again discusses the Dao together with language, and it claims that on their own, without discrimination or the deliberative process of intentionality, both the Dao and language are spontaneously what they are of themselves. In the state of nondiscrimination and non-intentionality, not only are things not taken out of their harmony with the Dao, but only in this way are things also completely present and harmonized spontaneously. The opposite case is illustrated with examples of concrete acts of discrimination and intentionality, all governed by the limits of what counts as "allowable" or "unallowable," that sunder the original harmony of things being as they are of themselves. Discriminations impose borders on the ability of things to be as they are of themselves, and this impedes the free movements of the Dao penetrating all things. When things are bordered, they are wrenched out of that unity and isolated apart from all other things; things are "ordered." When things are taken out of their harmony and ordered, they become pieces in the grand construction of manmade experience apart from natural spontaneity; things come to exist in "formation." Once things are placed in formation, their connection with the free movements of the Dao, the source of transformation and life itself, is lost; things are "destroyed." The Sage allows both assent and rejection to proceed spontaneously. Without discrimination and the imposition of intentional decisions to allow or not to allow judgments on things, the Sage allows all alternatives to exist freely by simply allowing them to be what they are.

The final lines of this passage reveal some points of connection with the *Laozi*. Where the *Qiwulun* speaks of the Sage's harmonization of assent and rejection, called the "Double Walk" 兩行 (*liang xing*) because it allows both to proceed without discrimination, *Laozi* 27 speaks of the Sage who rejects nothing: "Therefore, the Sage is constantly good at saving humans and never rejects anyone; and with things, he never rejects useful things. This is called the 'Double Brightness' 襲明 (*xi ming*)."[36] The similarity in both diction and idea in these two passages again expresses the discursive congruity between these two writings.

What follows in the *Qiwulun* is the most radical exposition of the theme of the origination and dissemination of borders within the world that is found in the *Zhuangzi* writings as a whole. This section is presented within a structure framed around the early Daoist cosmology and vision of the world first presented in *Laozi* 42. This is the *Qiwulun*'s most complete rendering of the original, first-order harmony between humans, the world, and the Dao. It is immediately followed by a depiction of the disruptions caused to the free movements of the Dao and the breakdown of that harmony, due to the intentional agency of humans. The section closes by describing the soteriological potentialities of the Sage and his restoration of a deeper harmony between human existence, the world, and the Dao, in the inception of a second-order harmony. This section begins with the following:

> As for those of old, there existed in their understanding that which arrived at the ultimate. What about this ultimate? There were those who took things as not yet beginning to exist: this is the ultimate, this is the exhaustive, and it could not be added to. The next took things as existing, but there had not yet begun to be borders between

> them. The next took borders between things as existing, but there had not yet begun to be the distinction between assent and rejection. The displaying of the distinction between assent and rejection is that which brings about the deficiency of the Dao. That which brings about the deficiency of the Dao, is that which brings about the completion of love. Is there really then completion with deficiency? Is there really then no completion with deficiency? There is completion with deficiency.[37]

This passage offers an odd examination of the history of human existence in the world, which it posits in three stages: in the first stage, human beings are so in harmony with the Dao that they have no awareness of things as existing independently; this stage represents the primordial beginnings of the formation of the human world. In the second stage, human beings are aware that things do exist, and this already marks a fall away from ultimate harmony. The beginning of language and the use of words to refer to things, thus rendering things sufficiently clarified and outstanding, set off apart from the unity in which there was no awareness of independent and signifiable existence that names can be used for them, characterize this stage. In the third stage, human beings, aware of signifiable things as independently existing, insert borders between them, thus wrenching things once and for all out of the flux and flow of the now reduced harmony. This stage witnesses the production of distinctions, which emerge in the form of genuine and artificial (true and false), and assent and rejection (yes and no). From this situation, human beings actively chop up reality through the application of judgments applied to things, including beautiful and ugly, right and wrong, and existing and not existing, all of which themselves are subject to being deemed allowable or not allowable. This explains the deficiency of the Dao, its darkening; its free movement and full presence in the world is disrupted, distorted, and impeded.

The *Qiwulun* attributes the responsibility for all of this to the intentional behavior of human beings, and its discussion of the completion of love is directly reminiscent of the ideas presented in *Laozi* 38. When human beings have reached such a state of depravity that human action is no longer performed through non-intentionality and the spontaneous expression of care, but rather is dictated by intentional awareness and behavior, then these acts are performed according to the dictates of intentional, artificial mores and values. They are performed with the intention of fulfilling specified criteria of good behavior, and the *Qiwulun* raises the criterion of "love" 愛 (*ai*). To fulfill the criterion of proper conduct and behavior is indeed to attain a completion, but this is "a completion with deficiency" 有成與虧 (*you cheng yu kui*), what the *Qiwulun* in an earlier passage called "a small completion"; these completions refer to any act performed by a human that successfully fulfills the criteria intended by that act. Opposed to these deficient completions is the great completion of the second-order harmony of the world effectuated by the Sage. This application of the notion of "completion" squarely places the *Qiwulun* within the early Daoist discourse of the period.

The *Qiwulun* goes on to give a much fuller exposition of each of these three stages by concentrating on the transitional moments leading from one to the next. The following passage describes the transition from the first to the second stage; the referent for the "this" is the first-order harmony which, by virtue of uniting all things as one, leaves no room for language or thought to posit a "not this." If everything is united, there can be no language to refer to separate things.

> At present there are indeed words about "this," but I do not know if they are of the category of assent or if they are not of the category of assent. If what is of the same category together with what is not of the same category are both deemed as categories, then "this" is without difference from "that."[38]

The central idea that the *Qiwulun* introduces here is explicated in the following passage by way of a discussion concerning the language of that-which-is and that-which-is-not, and the relation between the two within the unfolding of the cosmology and the generation of the world.

> There is a state in which that-which-is begins.
> There is a state in which "the state of that-which-is begins" has not yet begun.
> There is a state in which "the state in which 'that-which-is begins' has not yet begun" itself has not yet begun.
> There is a state of that-which-is.
> There is a state of that-which-is-not.
> There is a state in which "the state of that-which-is-not" has not yet begun.
> There is a state in which "the state in which 'the state of that-which-is-not' has not yet begun" itself has not yet begun.
> Suddenly there is that-which-is-not and that-which-is, but we do not yet know the condition of this that-which-is and this that-which-is-not. Does it belong to that-which-is? Does it belong to that-which-is-not?[39]

In the closing lines of this passage, the *Qiwulun* questions whether what has just been said, namely the linguistic affirmations concerning both that-which-is and that-which-is-not, can truly refer to these two or not. Speech is of the category of that-which-is. But can speech, belonging to this category as something deemed to exist, speak that-which-is-not? In other words, can one, relying simply on that-which-is, appeal to and realize that-which-is-not? Do they belong to different categories of reality in an ontological sense, or do they really just belong to one single linguistic category falsely distinguished by words? Is the distinction between them anything more than words, or is there really a that-which-is-not that can be grasped and attained from the position of that-which-is? Is the passage from that-which-is to that-which-is-not a kind of metaphysical continuum, or does it require a fundamental transformation to go from one to the other?[40] This passage is not a dissertation concerning language, as many commentators believe. It is a fundamental questioning of the nature and connection of that-which-is and that-which-is-not, carried out in order to explore the basic nature of undivided unity and the possibility of attaining it. More specifically, the issue involves whether or not the distinction between that-which-is and that-which-is-not is of the same nature as the distinctions posited by intentional human behavior: are all distinctions the product of human intentionality, or is this distinction part of the very fabric of reality? If the distinction between that-which-is and that-which-is-not really is a part of the fabric of reality, is it so by itself or only because humans make it like that by the use of bordering language? Finally, if humans do border reality through the impositions of borders, does this really affect reality or only the human understanding of it?

This following passage shows how oppositional distinctions can be overcome by allowing all alternatives to proceed, and it attempts to do this by smashing through the limits of language signification in order go beyond and attain the realization of unity. Displaying a series of paradoxes reminiscent of the many paradoxes of the *Laozi*, it says: "Nothing in the world is bigger than the tip of an autumn hair and Mount Tai is small; nothing lives longer than a still-born infant and Pengzi died young; Heaven and Earth are born simultaneously with me, and I and the ten thousand things are one."[41] The order of these paradoxes is strategically given, and, by demonstrating the break down of language signification in the face of reality, they attempt to smash through the borders imposed by distinctions. In order, the passage affirms the oneness of space (autumn hair and Mount Tai), time (infant death and longevity), the world (Heaven, Earth, and me), and all beings (I and the ten thousand things). This order recapitulates the grand themes of world completion: the cosmology of the beginning, the ontology of the present, and the soteriology of completion.

The section continues by depicting the spontaneous environment of human beings present at that first stage, a stage, it will be remembered, that preceded language. In order to isolate and explain the transition from the first stage to the second, it demonstrates what it means for there to be language and referentiality. Immediately after having affirmed the complete oneness of beings ("I and the ten thousand things are one") the passage continues:

> And now that we already are one, am I yet able to put that-which-is into speech? And now that I already said "this" is one, am I yet able to put that-which-is-not into speech? This oneness together with

> the speech for it makes two; these two together with [that act of saying that these two are] one makes three. To continue on with this is something even the cleverest mathematician could not handle, how much less an ordinary person. Therefore, in moving between that-which-is-not and that-which-is we arrive at three, and how much worse it is as we move between things and things. Without moving forth, proceed by affirmation only![42]

The structure of *Qiwulun*'s explication exactly follows the cosmological sequence of *Laozi* 42, which moves from the Dao to the one, to the two, to the three, then to phenomenal multiplicity. Having noted this general structure, I will return to the interpretation of the passage itself. Because the first stage was devoid of discriminating language, people were able to maintain themselves together with the unified being of the world and the Dao within the perpetual interplay of that-which-is and that-which-is-not. Language did not exist with which to discriminate things, but once discriminative language initially begins to operate, the harmony of all beings has already lapsed into multiplicity and human beings are primed to start bordering. This is the primary characteristic of the second stage of the fall from ultimate harmony, which the *Qiwulun* discusses in the following passage. Demonstrating the origins of borders, it returns to its focus on the Dao and speech, which are, it should be pointed out again, the two topics found at the very beginning of *Laozi* 1.

> The Dao never had borders and speech never had constancy, but when there were intentional affirmations then boundaries were made present. Let me have a word about these boundaries: there is left and right; there is hierarchy and correctness; there is division and disputation; and there is contention and warfare; these are called the Eight Virtues.[43]

Originally, before intentional affirmation establishes boundaries, the Dao was not bordered but enjoyed complete freedom of movement. Speech as well was not completely fixed since all things shared in the harmony of the Dao and its spontaneous transformations. Because things also enjoyed freedom of movement, speech discriminations were unable to keep up with their spontaneous transformations. With the subsequent exercise of affirmation, things get pulled out of the natural flux of change and transformation, unable to return to that natural flux because of the power of language to mold beings in accordance with signification and referentiality. For example, to call a cup a cup is to make it difficult for the cup to be a bowl; the name constrains the ability of the thing to be or become anything else. Things are harnessed to names and kept within boundaries delimited by borders. Human beings now are able to manhandle the spontaneous movements of existence through the impositions of the distinctions of thought and language, and from this they are able to produce the structures of the human world.

The *Qiwulun* passage provides a vision of the origins of the human world based on division, dissection, and the imposition of spatial coordinates on the originally nondiscriminated natural world and the things in it. Instead of a more typical and expected discussion of the origin of the eight directions of the world, it substitutes the Eight Virtues. From the early division of simple left and right, in which one side is always given precedence, there arises the notion of the left 左 (*zuo*) and the right 右 (*you*), or the base inferior and the noble superior. These two-term binary oppositions then translate into the social world with the emergence of social hierarchy 倫 (*lun*) with its higher and lower classes, together with the proper ritual and social acts of correctness 義 (*yi*) belonging to each of the social classes. With the borders already fully in place in the human world, discrimination begins to be exercised regarding what is allowable and not allowable, and the certainty of what constitutes proper form has begun to slip as the social framework of the human world begins to react against the uncontrollable elements of reality. When questions and doubts begin to play a role in affairs, and because there are no absolutely correct responses to all issues, human beings divide 分 (*fen*) into different factions united with themselves against others holding to different perspectives or positions, and the divided factions have only disputation 辯 (*bian*) to rely on in order to settle differences.

The theme of disputation is given great attention, particularly in the closing sections of the *Qiwulun*. I will not discuss that here, but merely point out that the great representatives of disputation in early Daoist writings are always the Confucians and the Mohists, each of whom affirm what the other rejects. Yet the *Qiwulun* has little faith in the power of disputation to resolve differences, the text being written, of course, in the period of the "Warring States." And so following upon disputation, there is contention 競 (*jing*), and here one is reminded again of the words of *Laozi* 38: "When someone does not conform to proper ritual etiquette, others then angrily roll up their sleeves and force them." With these first intimations of social violence, the tide cannot be stemmed and humans are left in such a position that strife 爭 (*zheng*), bringing with it the loss of human life and the total breakdown of peaceful, harmonious existence in the world, becomes the automatic response to issues that have no easy solutions.

The way to restore the breakdown lies with the Sage who, through embodying and uniting with the Dao and thereby making it present in the world, dissolves borders by refraining at every moment from inserting them in the movement of being.

> What is outside the Six Directions [up, down, left, right, front and behind], the Sage preserves without discussion; what is inside the Six Directions, the Sage discusses without appraising. Concerning the records of the former kings and their management of the generations during the Spring and Autumn period [of the Western Zhou], the Sage appraises without disputation.[44]

At every step, the Sage refrains from going too far, and his holding back allows being to proceed without becoming bordered. Specifically, the way of the Sage is to act non-intentionally; whenever intentional behavior arises in temptation to divide, dispute, or cut something by assenting to one part and rejecting the other, the Sage quiets it through maintaining an inclusive awareness that lets both parts proceed. "Therefore, with every division there remains something not divided; with every disputation, there remains something not disputed. You ask, 'How is this?' The Sage embraces it whereas common people dispute it in order to show it forth. Therefore it is said, 'Disputation leaves something not seen.' "[45] That "something not seen" is the undivided ground of the world that makes all division and disputation possible in the first place.

In order to express more fully the differences between fullness and borders, the *Qiwulun* compares what it calls "round" 园 (*yuan*) virtues to "edged" 方 (*fang*) virtues; the structure and ideas of this passage are again very similar to the opening lines of *Laozi* 38. There, the *Laozi* discussed the differences between "highest *de*" and "lowest *de*" and presented the genealogy leading from Dao to *de*, to humaneness, to righteousness, and finally to ritual. The *Qiwulun* also presents a kind of genealogy of five slightly different virtues, but instead of showing their line of devolution from the Dao, it marks the horizontal differences between "round" and "edged."

Huge Dao is not declared,
Huge Dispute is silent,
Huge Humaneness is not humane,
Huge Modesty is not humble,
Huge Courage does not hurt.

Dao brought forth is not Dao,
Words disputed do not reach,
Humaness made a norm is not complete,
Modesty made clear is not trustworthy,
Courage made injurious is not complete.

These first five are round (*yuan*) but soon develop edges (*fang*).[46]

The distinction between round and edged virtues is not the same type of distinction as between "this" and "that," or "self" and "other," but rather is the type of distinction that differentiates "great completions" (as round) from "small completions" (as edged). The round virtues characterize the non-intentional behavior of the Sage who restores the harmony of the world, and the square virtues characterize the intentional, bordering behavior of those like the Confucians and Mohists who labor to divide, order, and dispute.

As pointed out, the *Qiwulun* discusses the three stages of the origins, growth, and development of the realm of the Human from an environment of original harmony to an environment of bordered being. The final parts of this section examine the third stage, in which the ontological environment is already fully bordered, language signification is completely embedded in distinctions, and humans are not in harmony with the world. Its discussion of this final stage progresses by a series of dialogues among a variety of characters, which in itself is very telling. The dialogues are devoted to a series of discussions about the Sage, and they offer several anecdotes concerning the inclusive attitudes and activities of the Sage. What these discussions imply by their content and format is the soteriological possibility of a transformation of the depraved third stage, characterized by borders inserted between things, that will result in a new state of complete harmony, the second-order harmony capable of completing the world.

In this section, the *Qiwulun* introduces a new term for the Sage, the Arrived Person 至人 (*zhiren*). My hypothesis is that this person is said to have "arrived" in that he has arrived at a state in which all acts are performed non-intentionally. The discussion introducing this figure concerns the attempt by a character named Ni Que 齧缺 (Chew Lack) to assert that there really is some ultimate ground on which agreement and affirmation can be established. He makes this attempt through a conversation with a character named Wang Ni 王倪 (Kingly Horizon). Ni Que asks, "Would you know anything that all things could universally affirm?"[47] Wang Ni answers in the negative by saying that different things appropriately live in different dwellings, have different diets, and have different understandings of beauty, with the conclusion that, "From my point of view, the doctrines of humaneness and righteousness, the paths of affirmation and negation, are inextricably tangled up; how would I know their distinctions!"[48] Ni Que then asks about a more basic distinction that he believes is clearly apparent—namely, concerning benefit and harm—insinuating, ultimately, that there is a fundamental distinction all things could affirm, namely that between life and death. But, Wang Ni explains, to the Arrived Person even this distinction is void.

> The Arrived Person is spirit! When the huge marshland burns he is not scorched, when the Yellow River and the Han River ice over he is not frozen, when rapid lightning splits the mountain and the winds rattle the oceans he is not startled. One like this drives a chariot on the breath of clouds, and rides the sun and moon, roaming beyond the four seas. Since death and life have no effect on him, how much less do benefit and harm?"[49]

Ni Que expresses, through his questions, the bordered, splintered nature of this third stage of the human world. He intends to initiate a dispute with Wang Ni in which they will argue what can and cannot be affirmed. Ni Que appears as a total participant in the discriminated world of distinctions, and he tries to think through it from the inside. Wang Ni will have no part of it, because he sees the deficiencies of assent and rejection, and he knows that once he begins to speak that language he will have no escape from it. So he goes beyond the terms of the argument by adopting the attitude of the Sage, who overcomes the limitations of all distinctions.

The final passage from this section of the *Qiwulun* culminates in a description of the early Daoist Sage. "Befuddled and dull" 滉芚 (*yutun*) are high compliments in early Daoist writings. Note that the rhetoric of the Sage forming a triad 參 (*can*) with Heaven and Earth is again brought up.

> Roaming astride the sun and moon, upholding time and space and serving as their conjunctive unity, he establishes them, though they are random and slippery. By being attached to time and space, we are honored. The multitudes of people slave, slave; the Sage is befuddled and dull. The triad is aligned for ten thousand years as one, completed, and simple. The ten thousand things being completely themselves, through this we mutually intertwine.[50]

This ability to pass freely from one realm to another, at any point in Heaven or Earth, is a freedom of movement that the Sage shares with the Dao. He unites the realms of Heaven, Earth, and the Human, both within his breast and in the world. The final lines celebrate the achievement of the second-order harmony, and give a tantalizing description of the soteriological return. What this passage does not explicitly give the reader is an indication of the transformed body of the Sage. This is the theme of the next chapter.

Chapter Five

Early Daoism and Soteriology

THE HEALED WORLD

My use of the term "soteriology" to designate one major domain of early Daoist discourse will most likely strike some readers as problematic. Its application within Christian theological discourse identifies it with notions of original sin, redemption, and eternity, but such ideas are completely alien to the writings at hand. Leaving aside discussions of the ultimate metaphysical truth of individuals, what I intend to capture most of all by using this term is the early Daoist concept of return, but not a return of the individual to its creator. Using the term "soteriology" to name this domain, I do not intend to call up the notion of "savior" (*soteros*), but rather the notion of "salvation as process" (*soteria*) based on the idea of separation and reunion or return. The return that lies at the heart of the early Daoist soteriology is thoroughly intertwined with their visions of the body and its inherence in the world.

Early Daoist writings employ several different graphs for the body: *xing* 形, *shen* 身, and *ti* 體. *Xing* designates the form of the human body, and its range of signification is limited to the material package of the constituent parts of the body as a whole. The terms *shen* and *ti* also designate this material package, but their range of signification also includes notions that are often best translated as "person" in English. They refer both to the physical, fleshly, and manifest body as well as to the experiential identity associated with the bodily self. In English, the sense of the term "body" is only tangentially related to the sense of the term "person"; in early Daoist discourse, the terms *shen* and *ti* do not clearly differentiate the body from the person, primarily because they share the same physical source. Both the body and the mind are products manifesting from the temporary configuration of different cosmic energies.

Personal identity first and foremost is intensely physical and thoroughly fleshly. The personal identity connoted by *shen* and *ti* is more limited than the meaning conveyed by the English term "identity"; in the early Daoist writings, it is understood in a specifically spatial sense that demarcates the physical presence of an individual body. The manifest body is never entirely containable within the space it inhabits; it is not a closed organism but eminently open to the natural and cosmic worlds. More exactly, a healthy and properly functioning body is one that is not closed off from the natural and cosmic worlds, but is closed off from, or protected against, the debilitating influences and effects of the social world; an impaired body is one that is closed off from the natural and cosmic worlds, but not from the social world. Once a body is closed off from the natural and cosmic worlds, its openness to the vitalizing energies received from them is destroyed and the body suffers deleterious effects similar to a plant uprooted from the earth. This results in a psychic self-containment and the consequent formation of a constructed self that is solidified by the physical, social, and political processes of psychic and mental stagnation. These processes themselves are the products of false notions of self-identification and self-possession, and create a vicious circle that the early Daoist writings attempt to dismantle.

Early Daoist discourse envisions the world as consisting of three related realms, Heaven, Earth, and the Human. The early relationship of these realms is often represented by family metaphors, in which the Human is considered the child of Heaven and Earth as father and mother; at other times it is described as a complete unity. Intentional human activity alters and damages the harmonious relationship of these three realms, and the founding of kingship marks the moment of the beginning of this breakdown: the purposive policies and actions of the King are the complete opposite of the non-interfering movements of the pristine Dao. The founding of kingship creates the political body, and requires that the body of an individual be closed off to the natural world in order to be oriented to the one tangible center represented by the King himself. Kingship also creates the family body as a necessary concomitant of the political body put in service to it: as the young body incorporates the habits and postures of obligation to the parents, one proves oneself qualified for full participation in the state. The realm of the Human changes into a self-subsistent and autonomous realm, in large part due to the establishment of the political and family bodies, and this development has serious consequences for the harmonious unity characterizing the original relationship of the three realms of the world. The founding of the bodily orders of human civilization entails a critical disruption that splinters the wholeness of the world and human existence within it. The founding of kingship translates into the production of both the family body and the political body, and the early Daoist writers weave these themes into their own visions of the past and present causes for the disorders of the body and the world.

These early Daoist notions can be compared to a very different set of notions informing early Confucian discourse. For that discourse, the physical body belongs first and foremost to the biological mother and father. Stating that one's body is the property of the biological parents maybe does not go far enough, because it ultimately is seen to be the property of the ancestors. One has a body in order to serve one's parents throughout their life by acts of filial piety, and also to serve one's ancestors, including the biological parents after they have passed away, by acts of sacrifice. To carry out one's dutiful service, the body must be maintained intact, and thus there are stringent moral and practical obligations demanding that an individual care for the body that has been entrusted to him or her. Mismanaging the body by allowing it to be damaged in one way or another has a tremendous significance for one's ability to properly execute one's ritual duties to the living parents and the ancestral line.

Asserting the importance of the bloodline minimizes the body's openness to the world and society at large in favor of the ancestral line. However, this depiction of the early Confucian perspective does not go far enough, because it disregards the political component that lies at its heart. In addition to belonging to one's ancestral line, one's body also belongs to the King, who also is commonly represented as the father who gives one life and to whom one is obligated for just that reason. In a fundamental way, the King makes life possible for his subjects: on a mundane level, this is so because he safeguards the proper progression of the seasons, the proper performance of cultured, virtuous behavior, and the proper means whereby human beings become human through the familial relationships. On a more complex level, the King is identified as Heaven, the supreme moral authority of the cosmos itself, although this aspect of the King recognized as father is couched in his designation as the "Son of Heaven." The King is sacred in the most fundamental sense, because his presence endows life to all individuals by allowing the moral force of Heaven to be present in the world. Without the King, human existence would be cut off from Heaven, thus leading to the immediate breakdown of civilized institutions, rendering human beings no longer human but merely beasts or barbarians. Thus, in addition to the parents and the ancestral line, the body also and ultimately belongs to the King; the debts incurred by having bodily life are properly discharged only through filial service to the parents, ritual service to the ancestors, and loyal service to the King.

In contrast, early Daoist discourse recognizes neither one's biological parents nor the political parent as having endowed one's body. Early Daoist writings radically overturn the political metaphors with their claims that one's true parents are Heaven and Earth, in which reference to Heaven is denuded of all associations with the King. Heaven and Earth, moreover, are not considered the parents of the human body alone, but of the bodies of all phenomenal beings. The human body is not different from the body of phenomenal reality, and, together with all other human and phenomenal bodies, it ultimately exists as one body born from Heaven and Earth. In this way, the notion of the body open to the world begins to take on a deeper significance.

To state that the individual is not obligated to the biological parents or to the King, however, does not translate into a debtless state of bodily existence whereby one can do as one pleases with the body; on the contrary, the

Daoist writers claim that there are stringent obligations that must be carried out, but no moral authority coerces their execution. These obligations perhaps are better named as imperatives for a certain way of attending to the body that direct a person to the fulfillment of its inherent potential; the coercion brought to bear for not attending to the body in the manners specified is simply the threat of an early death. The potential inherent to the body born from Heaven and Earth is fully present from the moment of birth: Heaven endows the body with *yang*, which is sometimes imaged in the form of *qi*, breath; Earth endows the body with *yin*, which is sometimes imaged in the form of *xing*, bodily form. The coming together of breath and form supplies the very definition of human corporeal existence. Heaven and Earth themselves, moreover, are the offspring of *yin-yang*: Heaven is born with the congealing of *yang*, and Earth is born with the congealing of *yin*. The birth of *yin-yang*, in turn, directly issues from the pristine Dao and its original or cosmic *qi*. In this way, according to the Daoist writings, human beings indeed do participate in an ancestral line, but the ultimate ancestor is not the founder of one's bloodline, rather it is the Dao itself. Once again, we witness the overturning of some of the most vital elements of the early Confucian discourse.

The vital components of the human body do not consist of the lineage blood transmitted through biological propagation, but of the cosmic materials derived from *qi* and *yin-yang*. When the infant is born, it already is in full possession of the several different components necessary for existence as a human, including breath and form. Because it is subject to time and the processes of aging, coupled with the mechanisms of the socialization process under the observant regulation of the biological and political parents described in the early Confucian writings, awareness that Heaven and Earth are the true parents is difficult to realize. Not aware of this, the body closes off from the natural and cosmic worlds, and this damages its ability to continuously absorb the cosmic components due simply to the failure to attend to them. This causes the energies of the body to wither away through depletion, and they ultimately return to their source in the natural and cosmic worlds, and the body dies. Conversely, if a person attends to these cosmic components through different types of internal cultivation, they can rejuvenate to become overwhelmingly powerful within the body. One term that is often used to describe this phenomenon is *de* 德, power or circulation. When a body achieves the free circulation signified by *de*, the cosmic components of the body transform and merge with their ultimate ancestor, the Dao. This transformation of the body and its energies is often discussed with the rhetoric of longevity, long life, and other terms referring to the transformed body's ability to endure change and the passage of time without letting the energies of the body disperse.

The question of long life in relation to early Daoism is a point of contention for modern scholars. For early Daoist writers, long life never constitutes a separate and isolated soteriological goal in itself. What is important is the maintenance of the body's vitality, and this vitality is recognized as being able to extend one's lifespan. Only a body at the height of vitality could embody the Dao; the addition of the Dao to an already vital body was more than enough to endow extreme longevity. Remarks about longevity appear in the writings only as an afterthought. Undertaking the quest for the Dao, the body stands as the first object of concern, and proper cultivation of its internal components received from Heaven and Earth is the only way to prepare the body for its hoped-for embodiment of the pristine Dao. The ultimate consequence of the physical embodiment of the Dao for the body is that it will remain perpetually vital while maintaining the integrity of its inherent components through continuous acts of transformation. The possibility of transformation is a quality inherent to the body precisely because it is ideally open to the natural and cosmic worlds and, beyond that, to the pristine Dao. Commonly, metaphors that describe the transformed body revolve around images depicting the complete access to all regions of the natural and cosmic worlds enjoyed by the Sage.

The longevity sometimes mentioned in early Daoist writings does not describe the perfected physical body of a human that is identical in every way with every other human body except that it does not age, but rather the transformed body that incessantly transforms and fuses with the forms of nature. The integrity of that body is not defined by having four limbs and a head; rather, it is defined by the holding together of the cosmic energies within any bodily form enjoyed by the Sage or Genuine Persons. *Huainanzi* 7 presents a fairly typical depiction of his transformed body of the Perfected Person:

> Although they are present they appear absent; although living they seem dead. They pass through what has no space. Ghosts are in their labor and spirits are in their service. They dissipate into the unfathomable and enter into what has no space. They constantly relinquish and exchange their shapes, and their ends and beginnings are like an endless round, and no one can grasp their principle of relations.[1]

Note that the *Huainanzi* ascribes a most pointedly creative meaning to "the principle of relations" (*lun* 倫), a term otherwise thoroughly entrenched in the Confucian discourse referring to the hierarchical ordering of human interaction.

Bryan Turner's discussion of the meaning and role of the body in modern debates about it can shed light on early Daoist and Confucian notions of the body. Turner writes that "ontologies of the body tend to bifurcate around foundationalism and antifoundationalism: is the fundamental nature of the body produced by social processes, in which case the body is not a unitary or universal phenomenon, or is the body an organic reality which exists independently of its social representation?"[2] The foundationalist position seeks to underscore the "common social processes . . . related to the conception, gestation, birth, development, death and disintegration of the human body."[3] The antifoundationalist position deals "with the social production of the body, with the social representation and discourse of the body, with the social history of the body, and finally with the complex interaction among body, society, and culture."[4] The underlying views of the body presented in the discourses of early Daoism and Confucianism in many ways anticipate the modern debates that cohere around the two positions outlined by Turner. For the early Daoist discourse, the various components of the physical body were seen to be given completely at birth; they can be cultivated or dissipated as the case may be, but anything added on to the body from the moment of birth ever after—for example, good judgment or a refined taste—work only to dissipate the original constitution, thereby destroying its initial integrity. In this sense, then, the Daoist body (which I will also call the foundational body) can be seen to share in the set of notions Turner identifies with the foundationalist position. For the early Confucian discourse, the body born at birth is in no way considered human; to make it so is the work of the parents and the state, at least at first, but the process is an affair that spans one's entire lifetime. Becoming human is not something for which one is personally responsible, because it was unthinkable that a body could be made human in isolation from the family and the state. In this sense, the Confucian body (which I will also call the constructed body) can be seen to share in the set of notions Turner identifies with the antifoundationalist position.

Much of what I have said thus far should make it clear that the early Daoist vision of the body can be understood only by seeing it in relation to the world in which it exists and achieves its full potential. Successful practice of internal cultivation is completely dependent on the energies that are made available to the body by its inherence in the world. The body depends on and is formed by the world by way of the cosmological process of the first-order harmony; the world, conversely, depends on and is completed by the transformed body by way of the soteriological process of the second-order harmony.

This preliminary examination underscores the following points. First, the body designates a central participant in the soteriology that culminates in a second-order harmony for the world and the longevity of one's body. Second, this soteriology is grounded in the possibility of the physical embodiment of the Dao in this very body of every human being; this is not to be understood by a philosophical anthropology based on theories of nature, emotion, or spirit, but by a clear understanding of the physical constitution of the human body. Further, a large portion of this chapter will examine the loss of this possibility for embodying the Dao—in other words, death. The early Daoist understanding of death, why it happens, how it can be retarded or avoided, and several related issues, is inseparable from its theories about the causes of the deterioration of the body, inseparable from their vision of the body. This, too, calls for a direct examination of the way the early Daoist discourse envisions the physical, fleshly, and manifest body.

The *Neiye* Describes the Body as *Jing*

The body envisioned by early Daoist discourse is foundational: the several fundamental components going into its formation are the same components active in the cosmological formation of the world, and consist of *qi* and *yin-yang*, which are provided directly from Heaven and Earth. The Dao offers the very facticity of bodily existence itself; *qi* endows breath, and manifests the modes of *yin-yang* in the body. The most common manifestations of *yin-yang* in the body are as *jing* 精, *qi* 氣, and *shen* 神. *Jing* refers to the basic vitality of the body and its hot fluids, including but not limited to blood and sperm. *Qi* refers to the vital breath that animates the body; the bodily possession of *qi* defines life, and the loss of it defines death. *Shen* refers to a very refined state of bodily energy as mind or spirit. All of these components are fundamentally material, but *jing* is denser than *qi*, and *qi* is denser than *shen*. The important point to keep in mind is that these terms all refer to the material components that constitute the physical body, which are the same components that also constitute the world.

Heaven and Earth fulfill the roles of father and mother in the formation of the body, and together they serve as vehicles for the direct transmission of the several components of the body's manifest physicality. Although this ascription is fundamental in early Daoist writings, there is no strict uniformity in designating exactly what the body receives from Heaven and exactly what it receives from Earth. In general, however, the body receives the denser components of its constitution from Earth, including its shape, form, and skeletal structure, with the more transparent components received from Heaven, including mind, spirit, and breath. Altogether these are nothing more than the different modes or manifestations of *yin-yang*; *yin-yang* in turn are the two modes of *qi*; and *qi*, finally, is simply the breath or vaporous condensation of the Dao. In these ways, Heaven and Earth are recognized as father and mother to the human body as offspring, while the Dao is recognized as the ultimate ancestor. The body, furthermore, is seen to be open to the world and to the Dao in a manner that is similar to the ways in which other early Chinese traditions describe the body as open to the bloodline lineage; for the early Daoists, the ultimate identification of the body potentially includes the entire natural world of Heaven and Earth as parents, and the cosmogonic Dao as ultimate ancestor. This is, strictly speaking, a genealogy of the body that represents a radical departure from the commonsense notions of biological genealogy.

The *Neiye* writings present what is probably the earliest of the early Daoist descriptions of the physical, fleshly, and manifest body.[5] Although there are certain elements of the *Neiye* that seem out of keeping with more mainstream ideas of early Daoist discourse, this is better explained by the fact that it represents a very early phase of that discourse than by identifying it as non-Daoist. The writings could be the product of some scholars of the Jixia Academy; complex theories about *jing*, *qi*, and *shen* became current in early

China in large part due to their development by several scholars associated with this academy.[6] I identify the *Neiye* as early Daoist primarily because it discusses these terms in relation to the pristine Dao, and it provides theories about their concrete applications to the physical body. For example, *Neiye* 12 briefly discusses the birth of the body as the merging together of a heavenly component and an earthly component: "For human births, Heaven supplies the *jing* and Earth supplies the body, and these come together to form the human. If they merge there is life, if they do not merge there is no life."[7]

The *Neiye* heavily emphasizes the role of the *jing* and the heart 心 (*xin*) and gives them an especially privileged relation with the pristine Dao that subsequent early Daoist writings do not adopt; for them, the Dao could not be so precisely localized in the body. *Neiye* 5, for example, states: "The Dao . . . is that by which the heart is cultivated and the body is aligned."[8] *Neiye* 4 states: "The Dao has no fixed place, but it will settle peacefully in a good heart. With a tranquil heart and regular *qi*, the Dao can be stopped."[9] Here, the *Neiye* clearly expresses its own understanding of how to embody the Dao: it is got by and in the heart. Further, while subsequent early Daoist writings give a more or less equal importance to the presence of the *qi* and the *jing* for the formation of the body, *Neiye* 1 states: "All beings are brought to birth only with *jing*."[10] In these and other instances, the *Neiye* gives a somewhat different role to *jing*, but I still consider these writings as early Daoist, albeit extremely early and therefore somewhat rough in comparison to later writings, rather than non-early Daoist.

The *Neiye* throughout focuses on theories of the human body, and it provides concrete methods for cultivating it in order to make it into an appropriate shelter for the Dao. Many sections are devoted to discussing the heart, because it is the central lodging place for the Dao, but if the body languishes, the condition of the heart is irrelevant: the Dao will never come. Once the body is cultivated, one can concentrate on the heart. For the Dao to enter and reside therein, it must be made empty of internal commotions caused by violent emotion and thought directed to the world. The emptiness of the heart is achieved and maintained by the cultivation of tranquility, whereby these commotions are controlled and silenced.

For a person to be able to silence the internal commotions, the physical body must already be well cultivated. The initial stages of bodily cultivation are discussed in terms of nurturing the *qi* to allow it to flow evenly throughout the body. Improper bodily habits, such as overeating or undereating, damage the body and create circulatory problems. The *Neiye* describes appropriate meditative exercises that center on keeping the body and the spine straight and aligned in order for the *qi* to circulate freely. *Neiye* 13 states: "The life of a human depends on the straightness and alignment [of the body]. That which causes this [life] to be lost is inevitably exuberance, anger, sadness, and vexation."[11] A program of internal cultivation, described in *Neiye* 10, can deal with the emotional disruptions to the tranquility of the body: "You must coil and contract; you must uncoil and expand. You must be set; you must be grounded. Maintain this excellent [discipline] without cessation. Expel the excessive and discard the superfluous [*qi*]. As you realize the extent of this, you will return to the Dao and the *de*."[12] The *Neiye* attends to the body time and again through precise predictions of the changes it will undergo; *Neiye* 9 states: "If a person is aligned and tranquil, the skin and flesh will be supple and lithe, the ears and eyes will be acutely perceptive; the muscles will be limber and the bones strong."[13] These changes to the body are the result of the adept's ability to manage and cultivate the essential components received at birth, and represent one's progress to embodying the Dao.

The concrete discussions of the physical body given in these preliminary sections of the *Neiye* remain the constant backdrop against which to read all other sections, and this concentration on the physical body is entirely in keeping with the general tenor of early Daoist discourse. The reason for this sustained focus on the body in all of the writings is due to the fact that the body provides the best and most immediate access to the pristine Dao: by embodying the pristine Dao, the body manifests it in the world, and this directly brings about the soteriological completion of the world. For the *Neiye*, the heart is the specific place where the Dao can lodge in the world, and to maintain the heart at such a high level of emptiness and readiness, the body must be functioning perfectly. *Neiye* 3 describes the relation of the Dao to the uncultivated body: "The Dao infuses (*chong* 充) the body, but people are incapable of controlling it: when it goes, it cannot be made to return; when it comes, it cannot be made to stay."[14] A cultivated body is prerequisite for achieving physical union with the Dao, but achieving that demands a long time and continuous effort. A healthy body is one in which the *qi* and the *jing* powerfully coarse and circulate; a dead or languishing body will never know the Dao. *Neiye* 11 and *Neiye* 12 speak of the need to maintain the body's holistic health; *Neiye* 11 states: "Thought and inquiry generate knowledge; inattention and carelessness generate sadness; violence and arrogance generate resentment; sorrow and melancholy generate sickness; sickness and trouble bring death. . . . If you are not set to prevent this, your life will relinquish its abode."[15] *Neiye* 12 states: "When the torso masters straightness and alignment, and this method is regulated in the heart, long-life will be the result."[16] These passages demonstrate the absolute centrality given to the body.

To embody the Dao is to lodge it in the heart. If the body languishes through being uncultivated, then circulation and respiration will be blocked and shallow, and the heart will be the constant victim of the internal commotions caused by violent and reactive emotions and mental distractions, making it impossible for the Dao to take its place therein. The heart must be made empty of these disturbances; *Neiye* 2 says: "The nature of the heart benefits from calm and serenity: do not agitate it, do not disrupt it, and its tranquility will perfect spontaneously."[17] *Neiye* 4 says: "The nature of the Dao is adverse to sound and noise, but cultivate the heart and tranquilize its noise, and the Dao will be grasped."[18]

The heart is made tranquil by vitalized *jing* (as blood) made to flow throughout the veins, arteries, and other bodily passages. It clears away internal disruptions and blockages, thereby allowing the *qi* (as breath) to circulate unimpeded throughout. The *Neiye* discusses a particular relationship between the *qi* and the *jing*. A person is responsible for vitalizing the *jing*; the vital *jing* for its part spontaneously vitalizes the *qi*. The vitalization of bodily *qi* empowers the entire body, thus rendering it capable of sustaining the emptiness of the heart that will allow the Dao to enter it. *Neiye* 7 describes the free circulation of *qi* in the body in terms of the *de* that is said to "arrive spontaneously in surges."[19] When the body is empowered by *de*, here maybe best understood as pure and powerful circulation, the *jing* undergoes a fundamental change by expanding throughout the body, causing it to radically transform. *Neiye* 8 describes this change of the *jing* with the metaphor of a gushing wellspring.

> When *jing* is present it spontaneously enhances life; on the outside one is calm and radiant, on the inside (*jing*) amasses like a wellspring. Like a flood it equalizes and levels, becoming a fount of *qi*. Since the fount never goes dry, the four limbs are firm. Since the wellspring never drains, the nine apertures are clear. You can then exhaust Heaven and Earth and spread over the four seas.[20]

This important passage describes the changes of the bodily components received at birth, *jing* and *qi*, which follow from successfully cultivating the body and maintaining the emptiness of the heart. These changes of *jing* and *qi* directly cause the transformation of the body, here presented in terms of clear sensory awareness and the ability to journey far off by having universal access to the world. The clearness ascribed to the senses does not refer to easy access, in and out, of the internal components through the eyes, mouth, and the other apertures of the body; on the contrary, one needs to seal off the body or, more specifically, as the text states, "the wellspring," to recirculate, and thus rejuvenate, the components in a closed-loop system.[21]

The *Neiye* also incorporates discussions of the *shen* in its theories of the body, which counts as one of the three primary components commonly named in early Daoist discourse to characterize the foundational body (together with the *jing* and the *qi*). The *shen* is the most rarefied of the body's components, associated with the mind or spirit. The *Neiye*, however, gives it only a secondary place of importance behind the *jing* and the *qi*. The *shen* accumulates knowledge, but the intentional pursuit of knowledge poses a definite threat to the well-being of the body as a whole, as described in *Neiye* 11 previously quoted. With the transformation of the body, the *shen* also changes from something like an ordinary mind with a limited capacity for knowledge into an extraordinary mind with an unlimited capacity for knowledge, as described in *Neiye* 7.

The *Neiye* consistently attends to the physical, fleshly body, and it names the *jing*, *shen*, and *qi* as its fundamental components. It stresses the idea that these components must be restored to their initial state of vitality; if not, then they gradually will deplete and render the body sick and dead. Their vitality depletes from physical, emotional, and mental disruptions that divert one's attention away from the body to the outside world. Attention directed to the outside world debilitates the body because it allows the vitality of the components to seep out instead of being recirculated within; the *Neiye* describes this as the drying up of the wellspring. This is especially critical in relation to the *jing*, which, for the *Neiye*, is the very stuff of life. Attending to the body allows the Dao to be embodied physically within the heart. There is no disembodied, mystical vision in the *Neiye*.[22] The heart, named as the physical seat of the Dao, cannot be detached from the body in the same way that the spirit spoken of by mystics can be set free, and this also holds for the *jing*, *qi*, and *shen*. The *Neiye*'s depictions of the physically embodied Dao are severely corporeal, and it is not something that can be achieved with an emaciated body practicing yogic austerities. These ideas are clearly stated in *Neiye* 15:

> The Dao will spontaneously come; you can depend on it and forge all activity. If you are tranquil you will obtain it, if you are agitated you will lose it. . . . If the heart can maintain tranquility, the Dao will spontaneously stabilize. For the person who obtains the Dao, the pores are effused with it, the hair is saturated with it, and within the chest cavity there is nothing lost.[23]

To bring out further the distinctive qualities of the *Neiye*'s vision of the body, I will briefly compare it to a very different early Confucian vision set forth by the *Mencius*. The writer Mencius was one of the many early Chinese thinkers who spent time at the Jixia Academy, and it appears that he also developed his own vision of the body based on the complex of notions centered on *jing*, *qi*, and *shen* that he encountered there. Discussions of *qi* do not have an important place in the early Confucian discourse, and to find the *Mencius* providing one is almost an anomaly. The discussion occurs in only one section, and nothing else in the text bearing his name is even remotely similar to it. Although it gives a certain priority to the *qi*, the body described by the *Mencius* has certain distinctive qualities that identify it with the main themes of early Confucian discourse, and that sharply distinguish it from early Daoist descriptions.

For the *Mencius*, the exemplary body is best represented by the "heart that is not agitated" 不動心 (*budong xin*).[24] On the surface, this seems to have much in common with the *Neiye*'s understanding of the heart that is tranquil, calm, and serene, but this is deceptive. The *Mencius* sets forth a method for achieving a heart that is not agitated based on the cultivation of moral courage rather than the cultivation of the foundational components. Moral courage is cultivated by the constant exercise of unswerving righteousness, doing the right thing at every moment. To constantly practice unswerving righteousness requires the heart to be clear about what constitutes righteousness, and also to possess the physical energy needed to perform it; the *Mencius* writes of this energy that "it is born from accumulated acts of righteousness and it cannot be gathered by sporadic acts of righteousness. When a person's conduct falls below the standard set by the heart, then it will collapse."[25]

The *Mencius* states that the *qi* supplies the physical source of the energy that is employed in doing the right thing. As the physical source for the energy of the body, the *qi* is thus the driving force that also manifests the strength of emotions, passions, and desires produced in the heart. A central element of the *Mencius*'s view of the body is the four sprouts 四端 (*siduan*) of virtue that are present in seed form in the heart of every human being; they are received at birth directly from Heaven, here representing the supreme moral authority of the cosmos. For the *Mencius*, a human being is not considered to have achieved true humanity until the four sprouts are nurtured and come to full growth within the character of the person. Its description of the body, then, is a description of the ways in which these four sprouts are brought to maturity. Although its initial discussion of the body begins by examining the physical constitution, it quickly turns to the cultivation of the moral structures of character. It is valuable to note the *Mencius*'s use of certain central notions characterizing the early Daoist view of the body that he became familiar with from the Jixia scholars because this will help to bring out the distinctive nature of the Daoist visions by way of counterpoint.[26]

The four sprouts of virtue, properly nurtured, mature into the virtues of humaneness 仁 (*ren*), righteousness 義 (*yi*), ritual 禮 (*li*), and wisdom 智 (*zhi*).[27] Present within the heart, they are nourished by the circulation of a strongly flowing *qi*. The circulation of *qi* is empowered by moral courage, the result of the constant exercise of unswerving righteousness. A powerfully circulating *qi* is reflected in proper moral behavior, in good health, and in a clear complexion. Up to this point, the *Mencius* appears to be discussing the strictly physical constitution of the body, claiming that moral behavior itself is a natural component of the physical structure. If a person does not practice moral courage but instead allows activity to be dictated by passions, emotions, and desires, then the circulation of the *qi* is blocked, and its energy depletes and leaks away, particularly during strong emotional outbursts of fear, anger, or exuberance. This loss of the energy of the *qi* gives rise to mental, physical, and moral disorders and perversions. Immoral behavior, however, is not caused by the *qi*, but rather is the consequence of a depleted *qi* that has no power to strengthen the moral resolve to do the right thing.

The *qi* gathers and becomes vital in the body through unswerving acts of righteousness; when acts of righteousness have been performed for an extended period of time, the force of the *qi* coursing through the body is like a powerful river, invigorating the body's organs, both internal and sensory. The "flood-like *qi*" 浩然之氣 (*haoran zhi qi*) is the moral source that nourishes the four sprouts and brings them to full development; this phrase seems to be quoted directly from the *Neiye*, but the *Mencius* gives it an entirely different meaning. At this point, the *Mencius* departs from the terms of early Daoist discourse; for it, the *qi* is made powerful not by cultivating the body, but by performing virtuous acts.

The *Mencius*'s program of moral cultivation indeed begins with a form of physical cultivation, but it ends with a program of moral cultivation. Initially controlling the dominant emotional dispositions through regulating the circulation of the *qi* is clearly a form of physical cultivation, and when these are brought under control, the depletion of *qi* is restricted. The *Mencius* writes that the *qi* itself is controlled by the intention 志 (*zhi*): "The intention is commander over the *qi*, and the *qi* is what fills the body. Wherever the intention arrives, the *qi* immediately follows. . . . The intention, when blocked, agitates the *qi*; the *qi*, when blocked, agitates the intention. Now frustration and haste excite the *qi*, and it turns back to agitate the heart (*dong xin*)."[28] The intention is the direction, even the agent, of thought within the heart; it is that which performs constant acts of judgment, deliberation, and decision-making concerning issues of right and wrong and good and bad. Thus, to control the *qi* and nurture the sprouts of virtue in the heart necessitates unceasing discrete acts of the moral will; each discrete act performed by the moral will further empowers the *qi* in the body, which supplies the physical and moral courage for one's determination to accumulate acts of righteousness. The *Mencius* here mentions the "flood-like *qi*": "I am good at nurturing the flood-like *qi* . . . This *qi* is vast and unyielding. When it is nurtured with correctness and is undamaged, it fills the space between Heaven and Earth. This *qi* unites righteousness with the Dao."[29] Whereas the *Neiye* states that the heart physically gets the Dao, the *Mencius* states that one's moral behavior in the world, specifically righteousness, is what unites with the Dao.

Two important consequences follow from the possession of "flood-like *qi*": first, the body has an unending source of energy for the performance of virtuous acts, and, second, the body has an unending source of energy for learning virtue and teaching it to others. The *Mencius* asserts this last benefit of the "flood-like *qi*" by quoting Confucius: "I study without respite and teach without getting tired."[30] These consequences are very different from the two soteriological goals posited in the early Daoist writings concerning the completed world and the longevous body; furthermore, the Daoist writings decry the expenditure of energy used in constant performances of moral acts that not only does not allow the *qi* to replenish itself, but also represents the most immediate way that human beings deplete the energy of the body's foundational components received from birth. It directly leads to the loss of those energies and early death.

The differences between the bodies of the *Mencius* and the *Neiye* are striking. The *Mencius* nowhere mentions the supreme goal of early Daoist discourse—namely, the physical embodiment of the Dao. The *Neiye* always returns to the physicality of the body, but the *Mencius* ultimately relies on the notion of the *qi* as a kind of substrate on which the moral structure of the person is constructed. This is further underscored by the *Neiye*'s emphasis on the heart as the corporeal center for the embodiment of the Dao; for the *Mencius*, the heart is important because it is the source of moral behavior.

Certainly the greatest difference between these two visions concerns the ideal state of the heart. For the *Neiye*, the true nature of the properly cultivated heart is tranquility, silence, and emptiness. The true nature of the heart described by the *Mencius* is anything but tranquil: it is constantly directed to the performance of discrete, intentional acts involving judgment, deliberation, and moral choice. The Confucian heart is an extremely bustling place where no rest is allowed. I am spending a little extra time examining the *Mencius*'s vision of the body in order to underscore the Confucian concern with the heart as the center of human moral activity; it is only through the moral cultivation of it that a person achieves true humanity in the fullest sense of the term. To be a Confucian demands an unrelenting commitment to the constant performance of what is deemed right. The exertion of energy directed to proper moral performances is precisely what the early Daoist writings claim will deplete the body of its vital energies, and if this is sustained over extended periods, the body will exhaust itself and die early. The *Mencius* nonetheless argues that by actively engaging the *qi* in these discrete acts of righteousness, it will somehow constantly replenish itself, but it offers no explanation of just how this might occur.

The *Mencius*'s discussion does not recognize the possibility of physically embodying the Dao, but only that one's moral activities in the world can accord with it. It does discuss the blossoming of the four sprouts into the four virtues, but this is not exactly the kind of radical transformation discussed in the early Daoist writings. The blossoming is based on the ability to know right from wrong and good from bad and act on it, but this knowledge itself is not a natural component of the human body, it must be learned. The educative process begins with the biological parents and the family, in an environment where one learns to control one's passions, desires, and emotions in accordance with the social rules of proper moral conduct. In spite of the *Mencius*'s efforts to claim that the moral structure of the human being is given in seed form at birth, its arguments were not universally adopted by other Confucians (at least not until the Song dynasty), much less the early Daoists. The *Xunzi*, after all, written after the *Mencius* and directly responding it, claims outright that human nature is inherently bad, and only through efforts dedicated to making it good can an individual complete the essential progress that culminates in one's achievement of proper humanity.

THE *LAOZI* DESCRIBES THE NEWBORN BODY

For early Daoist discourse, the physical, fleshly, and manifest body is foundational; for this view, as Turner writes, "The body (is) an organic reality which exists independently of its social representation."[31] The body is complete at birth, fully supplied with the essential components by which it enjoys a pure harmony with the natural environment. No change done to the body can add anything to its natural perfection, and the inculcation of learned body movements or postures, refined eating habits, or body ornaments or decoration, only result in disruptions to the proper functioning of the components and damage to them. The body is foundational also for life, understood as the active synthesis of the different components received at birth and contained in the body; death is their dispersal. Therefore, personal existence continues only so long as the body maintains possession of these components. Death occurs when these components seep out of the body and their vital potency is completely exhausted. The body is foundational because its manifest physicality represents the primary, initial, and ultimate object of attention, to the exclusion of objects, phenomena, and concerns that lie outside of it. These senses of the foundational body are clearly set forth in *Laozi* 44.

> Fame or health—which is dearer?
>
> Your health or possessions—which is worth more?
>
> Gain or loss—in which is there harm?
>
> If your desires are deep, you will have great expenditures.[32] If you store much away, you are bound to lose a great deal. Therefore, if you know contentment, you will not be disgraced. If you know when to stop, you will suffer no harm. And in this way you can last a very long time.[33]

Laozi 44 posits the body as the primary field of attention. In this case, the phrase "If you store much away" does not refer to the sealing up of the body's essential components; rather, coupled with the phrase "desires are deep," it refers to extraneous things, such as pride, that cause loss to them. What this passage does not present is another sense in which the body is held to be foundational, namely for the fulfillment of the soteriological goal, "completing the great project" (*cheng da shi*) or, as I have called it, the achievement of the second-order harmony of the world. This supreme goal is achieved when a person is able to locate and embody the Dao in his or her own body, in this very world. This achievement allows the Dao to be fully present in the world after it has been blocked out of one part of it, the realm of the Human. According to the ontology of the present state of the world, human intentional activities result in the establishment of borders within the being of the world, thereby disrupting and destroying the initial harmony enjoyed by the realms of Heaven-the Human-Earth. This situation causes a critical separation between humans, the natural world, and the pristine Dao, and threatens human existence and the existence of the world with a total separation from the powerful source of life itself, the Dao.

To restore the world's harmony, the Dao must be made present in it again, so that its life-sustaining energies and influences can heal the damage. This second-order harmony is not the same as the first-order harmony, primarily because of the active role played by the highest representative of the human race, called either the Sage or the Genuine Person. In this sense, the body is foundational because it stands as the locus for the physical embodiment of the Dao: the body provides access for it to become fully present in the world once again. Without the human body, the Dao would have no way to enter. Embodying the Dao in this manner brings completion to the world and endows the body with longevity.

The body is foundational in these three primary senses: for birth, for life, and for the soteriology. These are very much related to each other. First, the body at birth, exactly like the world, is the product of the cosmic elements born from the pristine Dao (*qi* and *yin-yang* in the form of *qi*, *jing*, and *shen*); second, their coming together in a body is life, their dispersal is death; and third, the body is the very site for the Dao to become fully present in the world. These ideas provide the physical, fleshly, and manifest body an important place in early Daoist discourse, and the *Laozi* gives them very sophisticated expression.

Many of its central themes are continuous with certain of those in the *Neiye*. The most important themes shared by the two writings include the view of the body as foundational in the three senses previously discussed; emphasis on directing one's attention to the processes occurring within it and not to the outside world; recognition of the threat posed to the body by letting the components seep out and deplete; presentation of methods to seal up and vitalize them; assertions of the importance of emptiness and tranquility; and depictions of physical embodiments of the Dao.

Unlike the *Neiye*, which gives almost exclusive importance to *jing* for having life, the *Laozi* sees it as the fluid principle of the body, which is inseparable, in the body, from *qi* as breath. *Laozi* 21 depicts the *jing* within the body of the pristine Dao in images that draw an analogy with an embryo's spermatic conception in the womb: "As for the nature of the Dao—it is shapeless and formless. Formless! Shapeless! Inside there are images. Shapeless! Formless! Inside there are things. Hidden! Obscure! Inside there is *jing*. This *jing* is absolutely perfect; inside it is true."[34] Here, *jing* represents not one of several components of the manifest body (as in the *jing-qi-shen* grouping), but rather the spermatic vitality of initial conception prior to any manifestation.

Laozi 42 depicts the cosmic origins of human birth, and shows that the human body comes fully endowed with all of the components essential for life. Instead of the *jing*, *Laozi* 42 appears to give a priority to the presence of the *qi* and *yin-yang* as constituting the essential components in the actual birth of the body: "The ten thousand things are held up by *yin*, and embrace *yang*; through the blending of *qi* they attain a state of harmony." This passage can be seen to describe the birth process: newborn infants are typically born face up, with earthly *yin* beneath them and heavenly *yang* above; within the body of the infant, *yin* mixes equally with *yang* and the *qi* or breath is harmonious. Human birth thus replicates the cosmic process of the formation of the realms of Heaven-the Human-Earth corresponding to the Three of the opening lines of *Laozi* 42. The newborn infant does not represent a helpless creature at the mercy of death and disease awaiting the proper care needed to stay alive; on the contrary, the newborn infant is the very model of the completely vital body, in which all of the foundational components enjoy untrammeled expression and circulation. In the wider discourse of early Daoism, the pure expression and circulation of these components is termed *de*; in these usages, *de* carries no moral connotation whatsoever: it is thoroughly corporeal. *Laozi* 55 pulls all of themes together in its description of the newborn infant:

> One who enjoys full circulation (*de*) can be compared to a newborn babe. Wasps and scorpions, snakes and vipers do not sting him; birds of prey and fierce beasts do not seize him. His bones and muscles are weak and pliant, yet his grasp is firm. He does not yet know the meeting of male and female, yet his penis is aroused—this is because his *jing* is at its height. He can scream all day, yet he will not become hoarse—this is because his harmony is at its height.[35]

The infant is the model of the perfect body because in it all of the many components freely circulate without hindrance. In this body, there are no obstructions or leakages, no wastage or depletion in the free flow of the components that would make them go around, under, through, or outside their proper course. Because they are not superfluously expended, they never exhaust, as they are directly connected to the Dao. In this passage, the male baby has an erection without sexual activity, and this is one expression of the coursing circulation that is sealed up in the closed loop of the perfect body. The body of the infant embodies the Dao, and the *de* flows through freely. This calls to mind the *Neiye*'s image of the wellspring; *Laozi* 4 identifies the unending source of this energy in similar terms, but does not localize the bottomless source within the heart: "The Dao is empty, but when you use it you need never fill it up again. Like an abyss!"[36] Directly plugged into the Dao, the body will never know depletion.

The image of the infant is sustained throughout the *Laozi*, where it is employed as the ideal model for the perfectly functioning body of the matured adult. *Laozi* 10 discusses this in relation to the *qi*, here understood as the breath of both the Dao and the body: "Concentrating your *qi* and making it soft: can you make it like that of a child?"[37] According to the sense of *Laozi* 28, the perfection of the body of the infant can be recovered when the *de* courses powerfully throughout the body: "When you know the male yet hold onto the female, you will be the ravine of the country. When you are the ravine of the country, your constant *de* will not leave. When your constant *de* does not leave, you will return to the state of the infant."[38] In this passage, the *Laozi* discusses "knowing the male yet holding onto the female"; its usage here is very similar to the usages of later writings that systematically designate the "male" and the "female" by *yin-yang*. Further, the *de*-circulation described in this passage is unobstructed, exactly like that of the infant.

The *Laozi* never explains what the *de* specifically consists of, but some respected modern interpretations understand it sometimes as the physical manifestation of the energy of the Dao localized in an individual being, and sometimes as the manifestation of virtue in the character of a person. For example,

Philip Ivanhoe writes that "for Laozi, *de* is the 'power' or 'virtue' that accrues to those who attain a peaceful, tenuous, and still state of mind."[39] He also writes: "In early Confucian writings, *de* took on a genuinely ethical sense . . . it came to denote what I call 'moral charisma.' "[40] I do not entirely agree with the way the term has been read in interpreting the early Daoist writings, because these modern readings underemphasize the radically physical level of signification intended in them. Understanding the physicality invested in this term makes apparent the difference between the early Daoist usages of *de* as "power" or circulation and the early Confucian usages of *de* as "moral charisma" or virtue. This difference already appears in the *Neiye*, which discusses the surging of the *de* in the body that is aligned and attentive to it. The *Mencius* also recognizes at least a partially physical signification to *de* that was related to the floodlike *qi*, although ultimately it uses this to undergird its notions of the moral development of the person.

Laozi 59 gives images of the *de*-circulation that surges through the body, and it states that this leads to the embodiment of the Dao, here called "the Mother of the state" 國之母 (*guozhimu*). This manner of referring to the Dao is not unusual; the Mother is a typical name used by the *Laozi* to describe the Dao; in this passage, the "state" should be read as referring to the empire (*tianxia*), which again is a common way for the *Laozi* to refer to the world. What *Laozi* 59 describes, then, is completely in keeping with the two soteriological goals consistently posited by early Daoist discourse—namely, the longevity of the body and the harmonization of the world.

> Early submission—this is called repeatedly accumulating *de*. If you repeatedly accumulate *de*, there is nothing you can't overcome. When there is nothing you can't overcome, nobody knows where it will end. When nobody knows where it will end, you can possess the state. And when you possess the Mother of the state, you can last a very long time. This is called having deep roots and a firm base, and it is the Dao of long life and long-vision.[41]

These sorts of references to the Dao as Mother indirectly portray the ideal human body as being just like the body of the infant; it enjoys a vitality experienced by everything newborn. *Laozi* 52 discusses the Dao in terms of "the Mother of the world" 天下母 (*tianxiamu*): "The world had a beginning, which can be considered the Mother of the world . . . Block up the holes, close the doors, and till the end of your life you will not be labored. Open the holes, meddle in affairs, and till the end of your life you will not be saved."[42] The description of the body in these passages also describes the condition of the fetus still in the womb. The relation between the fetus and the Mother is closed: the fetus receives nourishment, breath, and life directly from the body of the mother, as the cultivated human body receives nourishment, breath, and life directly from the Dao. When this relation between the fetus and the mother is broken, holes are opened and the foundational components of the body leak outside; for the matured body, this is primarily caused by diverting one's attention to the outside world to meddle in various affairs of social life. The primary value attributed to the body of the uterine infant is that it is closed in the sense that none of the bodily energies can leak out.

For the matured body, putting a stop to outside leakage is a central and immediate concern, because if the components leak out, their vitality will deplete and thus render the body inert, lifeless, sick, and dead. *De* designates the power of bodily circulation; ideally, this is to remain sealed up within, allowing it to replenish and rejuvenate. The early Daoist model of the body that allows its components to leak away is associated with the busybody attempting to change the world through redirecting the power of *de*-circulation outside into the world in moral performances, and often this is explicitly associated with the Confucian body.

In early Confucian discourse, *de* designates virtue or moral charisma. Proper *de* has to be learned from the family and individual study. Its expression in the world transforms not only the character of the one who possesses it, but also one's family as well as one's nation in ever-expanding circles of influence. In the early Daoist writings, this expression of virtue is extremely dangerous because of the energy expended in demonstrating one's possession of it. This theme plays a prominent role in the *Laozi*, as seen, for example, in *Laozi* 5: "Much learning means frequent exhaustion. That is not as good as holding on to the center."[43] *Laozi* 48 negatively compares scholars of book learning to those who follow the Dao: "Those who pursue study daily increase; those who listen to the Dao daily decrease. They decrease and decrease until they reach a point where they act non-intentionally. Although they act non-intentionally, nothing is not done."[44] Here, learning from books is contrasted with and opposed to listening to the Dao. Although there are many things to be learned from books, I imagine that what the *Laozi* has in mind are primarily proper forms of ritualized comportment, this most important of early Confucian concerns. Indeed, mastering the forms of ritualized comportment demands a program of cultivation for the body, but this kind of cultivation is thoroughly oriented to activity in the outside world. Listening to the Dao is an inner cultivation of the body. Although the rhetoric of "internal/external" is somewhat foreign to early Chinese discourse in general, early Daoist writings consistently encourage maximal attention to the foundational components of the body, while simultaneously they discourage relentless attention to outward behavior; it is in this sense that the *Laozi* denigrates the increase of knowledge of ritual forms and extols the decrease of the same.

To attend to the body and achieve union with the Dao necessitates that one's field of focus turn inward to concentrate the bodily energies within. The *Laozi* states that the loss of the essential vitalities is not simply due to the pursuit of virtue, but is the inevitable result for any person who attends to the outside world to the neglect and detriment of the body. In this regard *Laozi* 12 discusses the exhaustion of the body's inner vitalities and the ensuing early death by exhaustion. I have already examined the following passage in some detail in an earlier chapter, but a few additional comments are warranted.

> The five colors cause one's eyes to go blind. Racing horses and hunting cause one's mind to go mad. Goods that are hard to obtain pose an obstacle to one's travels. The five flavors confuse one's palate. The five tones cause one's ears to go deaf. Racing horses and hunting cause will drive a person mad. Therefore, in the order of the Sage: he is for the belly and not for the eyes. He rejects that and chooses this.[45]

The *Laozi* discusses the disruption of the body's integrity in terms of bodily holes that are not sealed up; these holes commonly refer to the nine openings of the human body, including the ears, eyes, nostrils, mouth, anus, and urethra. These openings allow leakages that are motivated by the seductions, pleasures, and excitement of the world that attracts one's attention outward away from the body, depleting its energy and leading to early death. The *Laozi* advises that the best way to defend against the seductions of the outside world is through the cultivation of tranquility and the weakening of desire. The cultivation of tranquility empties the body of anxiety and the desire to control the world. The primary physical consequence for the body is that the energies will not deplete, but stay vital through their embodiment of the Dao (described as an "abyss"), the source of renewal, replenishment, and rejuvenation. *Laozi* 22 writes: "Bent over, you will be preserved whole. When twisted, you will be upright. When hollowed out, you will be full. When worn out, you will be renewed."[46] These sorts of descriptions are typical of the Dao; in this passage, they describe the body in possession of the Dao. These ideas are pursued in *Laozi* 55: "When things reach their prime they get old; this is called not the Dao. What is not the Dao will come to an early end."[47] Finally, in *Laozi* 76, the theme of the longevity of the body, likened again to the newborn infant who is able to sustain its tranquility, is juxtaposed with the early death caused by the depletion and loss of the vital energies:

> When people are born they are supple and soft; when they die they end up stretched out firm and rigid. When the ten thousand things and grasses and trees are alive they are supple and pliant; when they are dead, they are withered and dried out. Therefore we say that the firm and rigid are companions of death, while the supple, the soft, the weak, and the delicate are companions of life.
>
> If a soldier is rigid, he will not win; if a tree is rigid, it will come to its end. Rigidity and power occupy the inferior position; suppleness, softness, weakness, and delicateness occupy the superior position.[48]

A most noteworthy aspect of the *Laozi* is its systematic concentration on the body as physical, fleshly, and manifest. The importance given to the body is largely due to the view that sees it as essentially constituted by the foundational components of the cosmos; born directly from the pristine Dao, these components naturally circulate without obstacles or leakage in the newborn infant. The perfect or genuine body of the matured adult is just like the body of the infant because both are devoid of the disruptions, blockages, and punctures that deplete the vitality of the foundational components. The *Laozi* thus consistently describes the genuine body as empty and tranquil. Genuine tranquility refers to the deep, unrestricted flow of *qi*; a steady and powerful *de*-circulation of the *jing*; and a heart that is devoid of anxiety, exuberance, and other extreme reactions experienced in response to outside circumstances. These violent emotions have deleterious consequences for the body causing irregularities that ultimately lead to an inevitable and early death. The physical embodiment of the Dao is impossible for the body that is incapable of containing it.

THE *ZHUANGZI* DESCRIBES THE BODY AS HEAVEN

The *Zhuangzi* shares the central complex of themes that cohere around the *Laozi*'s vision of the body. These include the view of bodily existence as a temporary gathering or synthesis of the different cosmic components, the necessity of sealing up the body so these components do not leak away, and the longevity that is achieved by cultivating them. *Zhuangzi* 22 supplies a most celebrated definition of bodily existence: "Human birth is caused by the gathering together of the *qi*. Gathered, there is birth; dispersed, there is death. . . . For this reason it is said: 'A single *qi* coarses through the world, and therefore the Sage values the One.' "[49] With this passage, *Zhuangzi* 22 appears to cite the cosmological progression of *Laozi* 42, and concretely identifies the *qi* as the foundation of the body's physical constitution. This *qi* is the breath of the cosmic Dao that breathes life into all things; it is also the undivided stuff that, through dividing into two, gives birth to *yin-yang*. Human birth is defined as the gathering together of the *qi* in a localized space, the body. This *qi* can be read as a general category that is made up of the different components from which the world itself is brought to birth, identical to the stuff of which humans are born. In this definition of life and death, we are presented with the complete collection of the different elements that, when held together in the body, constitute the bodily existence of humans.

As long as these components of the body are held together, there is life. They disperse because their vital energy easily depletes, and with depletion they no longer can be held together, rendering the body dead. If, on the other hand, these energies are kept vital and contained in the body, then the body will transform in union with the Dao. *Zhuangzi* 5 discusses notions about the necessity of closing the body in order to isolate and then cultivate the inherent components; it then immediately depicts the transformation that follows:

> Death and life, preservation and oblivion, failure and success, poverty and wealth, excellence and incompetence, slander and praise, hunger

> and thirst, cold and heat: these are the changes of circumstance, the processes of fate. Day and night they alternate right in front of us, and there is no way to know the measure of their origin. Consequently it is not worth it to allow them to disturb your harmony. Do not allow them to enter into the Numinous Storehouse. Maintain the flow of harmony and ease without losing them through the senses. Ensure that day and night there are no fissures, and then it will be the spring for all beings: this is to encounter and give birth to the seasons within one's own heart. By this, I mean the material is whole.[50]

In this passage, the *Zhuangzi* appears to coin a new phrase for the foundational body, the "Numinous Storehouse" 靈府 (*lingfu*). This image beautifully represents the extreme care and attention that the early Daoists gave to the foundational body. It is considered "whole" (*quan* 全) when its internal components, the "material" (*cai* 才), are not allowed to mix with the superfluous material not inherent to it that belongs to circumstance and fate. The transformation of the body is described by its identification with spring, the period when everything comes to life: the transformed body brings the world and all things to life. These themes are found throughout the *Zhuangzi*; *Zhuangzi* 23 presents them particularly poignantly in the words of Geng Sangzi: "People who maintain the wholeness of their body and their life store up their body. Keep your body whole, hold on to your life, and do not let your thinking go uncontrolled. If you practice this for three years, then you will master this teaching."[51] Geng Sangzi here explains the methods for keeping the body whole by using the term *quan* used in *Zhuangzi* 5; this term, besides denoting the sense of wholeness and completeness, also has the sense of a physical integrity in which no part of the body's innate constitution has been allowed to separate and leak off into the external world.

Zhuangzi 32 describes the body of the average human in the following terms:

> The knowledge of a common person does not go beyond the etiquette of appointments and courtesy. This wears out his *jing* and *shen*, and they become deformed and depleted. He desires to lead everybody to universal benefits, supreme unity, and empty bodies. Like this, he becomes dazed and confused in all times and places; his body tires and he will never know the supreme beginning.[52]

Note that the reason why the common person becomes worn out is not because the three benefits are in any sense negative (on the contrary, these benefits are of the highest good), but because that person desires to achieve them for all beings intentionally. *Zhuangzi* 32 also applies the same ideas about exhaustion in describing the ruler of a state: "His body is worn out by the concerns of the state; his wisdom is depleted by its affairs."[53] *Zhuangzi* 31 describes, in the words of an old fisherman regarding Confucius, the consequences for the body that result from the constant pursuit of virtue: "Certainly he is humane, but I fear he will not save his body from harm. The strain to his heart and the laboring of his body threatens his bodily perfection. Alas, how far has he separated from the Dao!"[54] *Zhuangzi* 3 underscores the very physical nature of maintaining a vital body: "Go along with the spinal artery and make it your path.[55] By doing so you can preserve your body, keep your life whole, nourish the family, and live out your years."[56] In this passage, the naming of "the family" might strike one as counterintuitive in Daoist discourse; what the writer likely has in mind is an extended definition of the family as the ten thousand things, an idea more in keeping with early Daoist discourse. The themes of longevity, resulting from the successful maintenance of the physical integrity of the body, and notions of early death by exhaustion, resulting from the depletion and loss of the vital energies due to a misguided concern with the external world, are pervasive in the *Zhuangzi*. I have specifically quoted these four passages because they single out the effects of bodily integrity versus physical exhaustion in relation to four types of people: the common person, the ruler, the person of virtue, and the person who cultivates the body. These passages concisely demonstrate one hugely important set of themes concerning the body that informs early Daoist discourse.

In many places the *Zhuangzi* relies on extended narrative episodes to present its own additions to early Daoist visions of the body. Perhaps the most interesting of these are in the dialogues between Confucius, set up as a kind of fall guy used as a sounding board for the presentation of early Daoist ideas in the *Zhuangzi*, and his student, Yan Hui. In *Zhuangzi* 4, Yan Hui is full of resolve as he informs Confucius of his intentions to go to a powerful state and visit the king who has been abusing his authority there; Yan Hui single-mindedly intends to reform the king's character and his policies. Confucius responds that Yan Hui must first of all attend to his own body before he can hope to reform other people. In his response, Confucius speaks of *de*. Again, in the discourse of early Daoism, *de* consistently refers to the powerfully circulating internal components of the body as a whole rather than to the outward acts of proper behavior. Confucius says:

> Now, the Dao does not desire commingling; comminglement leads to multiplicity, multiplicity leads to disturbances, and when people are disturbed, they cannot be recovered. The Arrived Ones of old first established it in themselves and only after that did they establish it in others. Until you have firmly established it in yourself, what free time do you have for the activities of tyrants? Moreover, don't you after all know that the *de* dissipates with the production of knowledge? The *de* dissipates with reputation, and knowledge emerges from competition. Reputations die prematurely, and knowledge is an instrument applied in competition. Both are instruments

> of violence, and neither can be applied to perfecting conduct. . . . To sustain the *yang* at its height puts one under great stress, the tension shows in one's face; common people do not understand this, so they suppress what another person agitates within them in order to preserve the tranquility within their hearts. What is called "the *de* that daily gushes forth" will never reach completion, let alone the supreme *de*![57]

In this passage Confucius distinguishes *de*, identified with the inner processes of the body, from knowledge, the product of an attention that is directed to the outside world, in this case the knowledge of how to persuade a ruler to reform. When one's attention is directed to the outside world, the body begins to suffer the ill effects of anxiety (caused, I imagine, by uncertainty about the true responsiveness of the ruler, the fear from the uncertainty of whether or not the ruler will kill you if he is unhappy with the persuasion, and so on) and *de* can no longer be recovered amid the internal upheavals caused by inner turmoil. Confucius continues his words to Yan Hui with a sustained discussion concerning the internal constitution of the body, and explains that when one continues in a state of high anxiety for extended periods of time, identified by a too heavy reliance on the internal bodily energy named as *yang* (and manifest in an increased bodily temperature that "shows in one's face"), the body can no longer compose itself in tranquility and will suffer serious consequences.

The dialogue continues as Yan Hui asks about the proper way to attend to his body, and Confucius tells him to carry out a "fasting of the heart" 心齋 (*xinzhai*). He explains that this manner of fasting has nothing to do with the methods of ritual purifications current in early China involving temporary abstinence from certain foods and drinks. It consists, rather, of a specific program of inner cultivation that will expel impurities (the purpose of fasting is, after all, to rid oneself of extraneous matter) and seal up and close off the inner components of the foundational body, This closing up of the body, Confucius says in the final lines, will result in a transformed body.

> Confucius said, "Concentrate your intention; do not listen for it with your ear but listen for it in your heart; rather than listening for it in your heart, listen for it with your *qi*. Listening stops at the ear, and the heart stops with the appropriate sign. As for the *qi*: it is empty and awaits arousal from things. Only the Dao can accumulate emptiness. This emptiness is the fasting of the heart." Yan Hui said, "When I have yet to succeed in being the agent, deeds derive from me; when I do succeed in being the agent, there is no me; can this be what you call the emptiness?" . . . Confucius said, "When there are no doors and no poisons (entering the body), then unify your house where you can lodge in the inevitable, and you will be almost there.

> ... If the channels through the ears and the eyes are internally cleared and knowledge is expelled through the heart, then even ghosts and spirits will come to dwell in you, not to mention all that is human! This is to transform with the ten thousand things."[58]

This episode is remarkable for a number of reasons. The general tenor of this program of inner cultivation has much in common with the methods of the *Neiye*, but the *Zhuangzi* passage does not give exclusive priority to the *jing* and the heart as the *Neiye* does. Instead of focusing on the *jing*, *Zhuangzi* 4 singles out the *qi* as the primary object of attention. Also, although the *Zhuangzi* calls this the "fasting of the heart," the method attends to the heart only as a preliminary concern and only insofar as the heart is the center of knowledge from which all knowledge must be evacuated. The passage describes the heart as the primary opening of the inner body to the outside world, and this must be sealed in order to be able to contain the *qi* without leakage. Once the heart is emptied of knowledge, then the program here described no longer mentions the heart, in contrast to the *Neiye*, but instead attends to the body as a whole, which in this passage is discussed in terms of a "house" 宅 (*zhai*). Metaphors of bodies as houses are, comparatively speaking, not entirely uncommon in religious discourse; compare uses of this metaphor in the early Buddhist *Dhammapada*, where the mind is likened to a house that ultimately will be renounced: "As rain does not penetrate the well-thatched dwelling/ So passion does not penetrate the well-tended mind."[59] The idea of the *Zhuangzi*, however, is that the house, if well secured and sealed, will not be discarded but instead endure. This image is very close to that of the "Numinous Storehouse" previously discussed. Differing from the *Neiye*, which emphasizes the heart as the specific lodging place in the body for the Dao, the *Zhuangzi* sees the entire body as possessing the Dao. In this sense, the body is more like a temple than a garage. But this temple continuously transforms.

Zhuangzi 4 states that "this is to transform with the ten thousand things" 是萬物之化也 (*shi wanwu zhi hua ye*). Early Daoist writings consistently associate bodily transformation with longevity, and this association has misled many scholars to interpret the *Zhuangzi* as saying that upon death, the energies of the body simply return to nature in dispersed form and that is all. A. C. Graham, for example, writes:

> In the exaltation with which Zhuangzi confronts death he seems to foresee the end of his individuality as an event which is both an obliteration and an opening out of consciousness ... It seems that for Zhuangzi the ultimate test is to be able to look directly at the facts of one's own physical decomposition without horror, to accept one's dissolution as part of the universal process of transformation.[60]

Zhuangzi 4, however, is saying something entirely different; by properly cultivating the components of the body, it transforms in such a way that it unites with nature; the components do not randomly disperse but continue to maintain their integrity within a body that constantly transforms. Witnessing the changes that the transforming body undergoes, one might not even know that it is the same body that a moment ago was identifiable by its human shape. This is the *Zhuangzi*'s particular view of longevity that is also a major defining feature of early Daoist discourse in general, and there are voluminous depictions of the constant transformations of Sages throughout the writings that bear this out.

The *Zhuangzi*'s understanding of the question of death versus transformation is actually somewhat logical. As *Zhuangzi* 23 claims, death is the complete dispersal of the internal components. For one who possesses a transformed the body, the components do not disperse; rather, they "transform together with the ten thousand beings." For the *Zhuangzi* and other early Daoist writings, longevity does not bring to mind the ability to sustain the shape and form of one's own human body throughout countless eons of time; rather, to achieve the transformation of the body is to attain the deepest identity of the body with the world. Longevity does not mean to keep the human shape of the body as given, but only to be able to contain the energies of the body and hold them together (and the later practices of inner alchemy would go on to mystify these teachings). To achieve the transformation of the body is completely to merge with nature while at the same time keeping the specific components of the body intact. Such an idea is presented in *Zhuangzi* 6: "If even after ten thousand transformations the body has not yet begun to reach exhaustion, then can its pleasures ever be calculated? For this reason the Sage makes present all [the vital components] and roams in the space where they cannot slip away."[61] The first step in transforming the body is to separate the foundational body from its involvement with the outside world and isolate, in order to directly cultivate, the inner components received at birth. This can be accomplished only by evacuating all other parts of the body and the person that are learned, constructed, or added on (this, indeed, is the goal of the "fasting of the heart"). These notions are again presented in *Zhuangzi* 6, and again in the medium of a discussion between Yan Hui and Confucius.

> Yan Hui said, "I made progress." Confucius said, "What do you mean?" Yan Hui said, "I have forgotten about humaneness and righteousness." Confucius said, "Good, but you have not yet arrived." Another day he saw Confucius and said, "I made progress." Confucius said, "What do you mean?" Yan Hui said, "I have forgotten about rites and music." Confucius said, "Good, but you have not yet arrived." Another day he saw Confucius and said, "I made progress." Confucius said, "What do you mean?" Yan Hui said, "I sit and forget." Confucius was taken aback and said, "What do you mean by sit and forget?" Yan Hui said, "My limbs and my body 體 (*ti*) fall away, my hearing and eyesight are put away, my body 形 (*xing*) departs and

> knowledge is expelled, and I unite with the great passage; this is what I call to sit and forget." Confucius said, "With such union there are no desires; in such transformation there is no constancy. Certainly you are worthy! I ask now to serve and to follow you."[62]

This passage thus represents a key rebuttal of a key word of Confucian discourse, namely *hao* 好 (nominally, desires; verbally, to love). In this passage, Confucius states: "With such union there are no desires (*hao*)." In the *Lunyu*, Confucius several times singles out Yan Hui for his "love of learning" 好學 (*hao xue*).[63] This passage, on the other hand, rejects the value of desires (*hao*) and the intentional notions of moral culture extolled by the Confucian discourse, namely humaneness (*ren*), righteousness (*yi*), ritual (*li*), and music (*yue*). Here, the early Daoist reversal of Confucian norms is at its sharpest and most explicit. Whereas Yan Hui is alleged by the tradition to be Confucius' favorite pupil precisely because he is said to love (*hao*) learning, this *Zhuangzi* episode depicts his attainment as one of eliminating this very desire.

Zhuangzi 11 explicitly links such kinds of desire to the exhaustion of the components of the foundational body. This discussion demonstrates an important trend in the early Daoist writings that first appears with the *Zhuangzi* and is taken up most noticeably by the *Huainanzi*, namely the application of *yin-yang* terminology to describe the proper balance (or imbalance as the case may be) of the bodily components that are either maintained by the lessening of desire or depleted by giving full rein to it.

> Are people too joyful? If so, then they harm the *yang*. Are people too angry? If so, then they harm the *yin*. If *yin-yang* are both harmed, then the four seasons will not follow each other and the balance of hot and cold will not be maintained, resulting by contrast in damage to one's very body. This will cause people to displace the proper place of joy and anger; they will be unable to stay in one place and their thoughts will not be able to concentrate on anything, breaking off midway without achieving any results.[64]

The *Zhuangzi* consistently contrasts the foundational body and its inner components to the constructed self and the artificial knowledge and the artificial desire that characterize it. Knowledge involves discrimination, judgment, reason, and cultivated taste, and desire refers to a developed emotional sensibility that produces strong reactions to circumstances in the outside world, such as anger, exuberance, and indignation. Artificial knowledge and desire in turn are constructed simultaneously with the psychic production of the constructed self in the form of deficient personal identities (see my discussion of this in the previous chapter). The *Zhuangzi* single-mindedly attacks any perspective that would value the artificial and constructed aspects of the self because their intentionality uses up as fuel the vitality of the inner components that ideally ought to be reserved for the needs of inner cultivation.

The *Zhuangzi* throughout prioritizes the foundational body at the expense of the artificial knowledge of the constructed self. These views are demonstrated in the episode with Nanguo Ziqi about the loss of his constructed self and the two dialogues between Yan Hui and Confucius about "the fasting of the heart" and the practice called "sit and forget." These three dialogues share a fundamental concern with sealing up the body, separating the inner components of the foundational body from the artificial knowledge of the constructed self in order to expel the latter. This renders the inner components of the foundational body free to circulate so that they can be cultivated directly in the pursuit of transformation without the obstructions created by the constructed self.

The *Zhuangzi* is a central and active participant in early Daoist discourse, as shown by the attention it gives to the foundational body, and it substantially furthers the limits previously explored by other writings. In this, the *Zhuangzi* does not fundamentally change the way that other early Daoist writings understand the body; instead, it reconfigures the different concerns going into those understandings and thereby establishes a more precise set of terms and images with which to discuss the foundational body in contradistinction to the constructed body. A passage in *Zhuangzi* 5 exposes the differences between the foundational body and the artificial aspects of the constructed body; this occurs in a dialogue between Zhuangzi and his companion, Hui Shi, in which Zhuangzi provides an explanation of what counts as foundational and what counts as constructed or artificial.

> Hui Shi asked Zhuangzi, "Can a human truly be devoid of [human] nature?"
>
> "He can."
>
> "Well, if a human is devoid of [human] nature, how can he be called human?"
>
> "The Dao provides the appearance and Heaven provides the body, so how can we not call him human?"
>
> "But since we do call him human, how can he be devoid of [human] nature?"
>
> "By [human] nature, I mean acts of assent and rejection. By saying that he is without [human] nature, I mean that he does not inwardly injure his body by likes and dislikes, and that he constantly adheres to spontaneity and adds nothing to his existence. Again, the Dao provides the appearance and Heaven provides the body; do not inwardly injure the body by likes and dislikes. But now you continue to push your *shen* outside [in pursuing knowledge] and exhaust your *jing*."[65]

In this dialogue, the term that I translate as "[human] nature" is *qing* 情, which I read as referring to the artificial aspects of a human being, specifically artificial knowledge and artificial desire.[66] Although many scholars commonly translate *qing* as "emotions" (Burton Watson uses "feelings," and Victor Mair uses "emotions" in their translations of this passage), Graham seems to be more on the right track with his translation of *qing* as "the essentials of man."[67]

There are two main points to my translation of *qing* as "nature." First, the *Zhuangzi* nowhere says that the Sage is without emotions, much less an ordinary person. The Sage, however, is without violent emotions, such as exuberance and rage; to be without emotions would be much closer to death than to life in Zhuangzi's view. Thus, in his initial statement Hui Shi asks not about the ordinary state of humans but about the original or authentic state. So *qing* must be something humans can be without and still be alive. Second, Zhuangzi specifically defines *qing* as "acts of assent and rejection" 是非 (*shi fei*) and "likes and dislikes" 好惡 (*hao wu*), and he furthermore claims that these "inwardly injure the body" 內傷其身 (*nei shang qi shen*). From what, then, do such phenomena emerge? According to the statement that humans are originally without these, they most certainly cannot be considered as part of the foundational body but, as Zhuangzi goes on to say, they are added on. More exactly, he says that the human in question (doubtless the Sage) does not add these on to his "existence" 生 (*sheng*). I expect that "existence" refers to the physical constitution of the body, odd though it sounds that "acts of assent and rejection" and "likes and dislikes" could be understood in any physical sense, but the result is, indeed, inner bodily injury. I think it is clear that for Zhuangzi, like many writers of the period, assent and rejection emerge from both an artificial or educated knowledge and an artificial or educated desire; these constitute what Zhuangzi means by *qing*. It should be noted that Zhuangzi completes his explanation of the foundational body by once again grounding it firmly in the context of its physical, fleshly, and manifest nature.

By discussing the differences between the foundational body and the constructed self through this use of *qing*, *Zhuangzi* 5 extends the limits of early Daoist discourse on the body. The *Zhuangzi* as a whole introduces a new set of terms and images applied as a kind of standard with which to understand the foundational body, rigorously distinguishing from it all that is not part of its spontaneous and original constitution, especially the artificial knowledge and the artificial desire of the constructed self. A further addition to the terminology applied in these discussions is the use of the term Heaven 天 (*tian*) to correspond to the inner components of the foundational body, and the term the Human 人 (*ren*) to correspond to the artificial elements of the constructed self. The significance of the *Zhuangzi*'s application of these terms is that it makes possible a way to express the radical differences between the foundational body and the constructed self, and to highlight that only the bodily components named as Heaven are essential for the cultivation of the body. The artificial aspects of the constructed self corresponding to the Human have only a negative value; naming them as the Human allows one to know them for what they are, and expel them.

The *Zhuangzi* also supplies a third term that shores up the notions expressed by his extended meanings of Heaven and the Human, namely *zhen* 真, literally "genuine," but signifying the perfected genuineness of the transformed body. As used by the *Zhuangzi* and other early Daoist writings, genuineness means the attainment of identity or union between the inner components of the foundational body and their cosmic sources, ultimately the pristine Dao. *Zhuangzi* 31 succinctly describes this genuineness:

> The genuine is what we receive from Heaven; it is spontaneous and cannot be replaced. For this reason, the Sage models Heaven and values the genuine, and he is not overcome by custom. The fool does the opposite. He cannot model Heaven and is anxious about the Human; he does not know how to value the genuine. Tediously he suffers the changes brought on by custom, and for this reason he is unsatisfied.[68]

This passage identifies the genuine with Heaven, and custom with the Human. The body from which everything corresponding to the term the Human has been expelled is left with nothing but what corresponds to the term Heaven; the person whose body is rendered thus is called the Genuine Human 真人 (*zhenren*). The application of these three terms, Heaven, the Human, and the Genuine in this way, appears original to the *Zhuangzi*, but the influence it exerted is seen in the fact that these three terms became a common staple of Daoist discourse from the time of the appearance of the *Zhuangzi*.

The following passage from *Zhuangzi* 6 provides the most succinct presentation of the different significations of Heaven and the Human in their application to the foundational body and the constructed self, respectively. The application of Heaven to the foundational body is resolved toward the end of this passage in a brief discussion of the transformed body of the Genuine Person.

> The person of achievement is one who knows the workings of Heaven and the workings of the Human. One who knows the workings of Heaven exists with Heaven. One who knows the workings of the Human uses what he knows through knowledge in order to nurture the knowledge of what is not known. The culmination of knowledge is to live out one's given years without dying early halfway through. . . . How do I know that what I call Heaven is not the Human, and that what I call the Human is not Heaven? However, first there are Genuine Humans and afterwards there is Genuine Knowledge. What is a Genuine Human? The Genuine Humans of old did not avoid being alone, they did not complete through virility, and they did not resent their circumstances. Such people as these did not regret their mistakes and were not arrogant

> from their successes. Such people as these climbed heights without trembling, they entered water without getting wet, and they walked through fire without getting burned. In this way, their knowledge ascended through falsity and attained the Dao. The Genuine Humans of old slept without dreaming, they awoke without anxiety, they ate simple foods, and their breathing was deep, deep. The breaths of the Genuine Humans came through their heels, while the breaths of ordinary people come from their throats. Dominated and submissive, ordinary people talk in gulps as if retching. In those whose desires and cravings are deep, the impulses of Heaven are shallow.[69]

Zhuangzi 5 continues these ideas with a sustained meditation on the radically physical signification of what is designated by Heaven in sharp contrast to the Human. This passage begins by stating that the Sage is directly nourished by his bodily Heaven and not by the Human, and its use of the Human is strongly reminiscent of Zhuangzi's definition of *qing* as "assent and rejection."

> Receiving his nourishment directly from Heaven, what use has he for the Human? He has the shape of a human and exists among humans, but he is devoid of the Human; therefore, assent and rejection have no place in his body. Whatever it is that identifies him as a human is indiscernibly small, while he is full of Heaven, and this he perfects in solitude.[70]

This passage exploits the distinction between Heaven as that from which the vitality of the inner components is nourished and which also completely fills the body, and the Human as that which is identified as assent and rejection—in other words, those artificial aspects of the constructed self defined as *qing*. Also notice that this passage explicitly differentiates the two senses of *ren* 人 as human being and the Human.

The essential force of the *Zhuangzi*'s application of the terms Heaven and the Human comes from its ability to have these terms refer to what I call the foundational body and the constructed self. This is a fundamental redefinition of these two terms in the total field of early Chinese discourse. The primary meaning of Heaven in other, non-early Daoist writings is simply the sky above, and in a derivative set of meanings it came to signify the supreme moral authority of the world. In early Confucian discourse, Heaven also is associated with the ruler of the state and the father of the family, and it represents a localization of the supreme moral authority that was not entirely removed from the world of human civilization. Exploiting this closing of the gap between Heaven and humans, the *Mencius* developed a moral philosophy concerning the cultivation of the heart, in which Heaven plants the seeds of virtue.[71] Through continuous moral cultivation, the seeds will blossom in the manifestation of correct virtue. The *Mencius*, however, identifies Heaven not with the physical body but with the moral structures received at birth by all humans.

The *Zhuangzi* takes an even more radical step by identifying Heaven with the collection of physical components received at birth, and it designates all later embellishments to the body as constructed additions that deplete those components. In the *Zhuangzi*'s case, Heaven serves as a category marker to name the inner components in opposition to the artificial structures that are created after the foundational body has already come together. "Genuine" describes the body in which those components called "Heaven" have transformed in union with the Dao. Another term that is used in a virtually identical sense with Genuine is *zhi* 至, literally "to arrive" or "arrival," and it signifies the arrival or attainment of bodily genuineness as something achieved by the cultivation of the body. *Zhuangzi* 15 uses this term in the following passage that demonstrates yet again the thoroughly physical signification intended by its use of these terms.

> A heart devoid of anxiety and pleasure has arrived *de*. Being single and unchanging is arrived tranquility. Having no obstacles is arrived emptiness. Having no intercourse with beings is arrived serenity. Rejecting nothing is arrived purity. It is said that if the body is overworked and allowed no rest, it will collapse, and if the *jing* is employed without cessation it becomes tired and eventually will exhaust. . . . Simplicity means that [the *jing*] is not diluted; purity means that the *shen* is not damaged. He who embodies purity and simplicity is called the Genuine Human.[72]

The programs that serve to keep genuine Heaven distanced from the artificial Human follow almost inevitably from the way in which the distinction of Heaven and the Human is set up: to preserve Heaven, one must first expel the Human; after expelling it, one must seal up the body so that the Human has no further access inside. Sealed away in the body, the collection of inner components designated as Heaven can be cultivated, and this results in the bodily manifestation of the *de* and bodily rejuvenation and longevity, the necessary prerequisites for embodying the Dao. These ideas are presented in a precise manner in *Zhuangzi* 19.

> One who penetrates the nature of life does not regard as duty what life does not act upon. One who penetrates the nature of fate does not aim to know what nothing can be done about. To nurture the body, one first has to manage things, and yet there are instances when there is more than enough yet the body is not nurtured. To possess life, one first of all must not alienate the body, and yet there are instances when the body is not alienated but one still dies. Life's coming cannot be resisted, its going cannot be stopped. Alas, people

> who are of the world believe that simply nurturing the body will be enough to preserve life. If nurturing the body is really not enough to preserve life, then what value is there in the worldly methods of nurturing the body? But even if they aren't worth doing, is there no avoiding what has to be done? Therefore, desiring to avoid concern for the body, nothing is more effective than abandoning the world. Abandoning the world, there is no more fatigue. Without fatigue, then one can be straight and aligned. Being straight and aligned, then there is renewal with what is other than oneself. With renewal, one is almost there. Is it worthwhile to abandon the world? Is it worthwhile to neglect that life? Abandoning the world, the body is not overworked; leaving behind the desires of life, the *jing* is unimpaired. One who has an intact body and who has returned *jing* becomes one with Heaven. Heaven and Earth are the father and mother of the ten thousand things. Their union forms bodies, their separation forms beginnings. When the body and the *jing* are unimpaired, it is called being able to move. From the *jing* to the *jing*, one returns to match Heaven.[73]

This passage serves as a general recapitulation of many of the central themes informing the *Zhuangzi*'s view of bodily existence. Many of its ideas are reminiscent of the early sections of the *Qiwulun,* but I won't discuss this in any detail. The first sections describe what is not essential for life—namely, concern with things that are beyond human control. What can be controlled is how one takes care of the body, and this is the prerequisite for continuing one's life. Nurturing the body, nonetheless, is no guarantee that a person will not die by some unforeseen circumstance, and this is simply fate. With that in mind, the passage states, the body still must be cultivated. The text then presents a series of preliminary tasks: to abandon the concern for the world and thereby contain the body's energies, rendering the body tireless and not overworked; and to direct that energy back into the body in cultivation practices, characterized as "straight and aligned" 正平 (*zheng ping*), terms also used by the *Neiye* to describe the initial stages of bodily cultivation. With a cultivated body, one "renews the body's energies with what is other than oneself" 與彼更生 (*yu bi geng sheng*), referring to the world consisting of the three realms of Heaven-the Human-Earth. The passage then speaks of the intact body 形全 (*xing quan*) and returned *jing* 精復 (*jing fu*), which are sometimes used as key words for *yin-yang*, and the union with Heaven, which is sometimes used as a key word for the Dao.

With this reading in mind, it can be seen that this passage from *Zhuangzi* 19 presents the central themes constituting the main ideas of the second-order harmony of the soteriological return. In addition to relying on the term Heaven to discuss the foundational body, its focus is thoroughly directed to the physical, fleshly, and manifest body, and the ideas and practices that will transform it. Transforming, the body follows the reverse cosmological order of *Laozi* 42: from the realm of the ten thousand things, in which the body finds itself subject to the ravages of time and mortality, the adept cultivates the foundational components and expels the superfluous. Rendered longevous and rejuvenated, the body unites with the world consisting of the three realms, and from there it returns to the realm of *yin-yang* and then to the union with the pristine Dao. This passage depicts the soteriological return only from the vantage of the body; it does not mention the consequences this return has for the completed world.

The great contributions made by the *Zhuangzi* in expanding the limits of early Daoist discourse include its development of the three terms, Heaven, Human, and the Genuine, and their application to the physical, fleshly, and manifest body. Heaven refers to the inner components of the foundational body, including the *yin-yang*, *jing*, *qi*, and *shen*. The Human refers to everything artificial that is added on to the foundational body, including the artificial knowledge and artificial desire of the constructed self. The third term, the Genuine, refers to the transformed state of the physical body that has cultivated the foundational components, and synonyms for this state include tranquility, emptiness, and harmony. One further addition made by the *Zhuangzi*'s discussions of the body is its application of *yin-yang* to describe the balance or imbalance of the inner components; bodily imbalance of *yin-yang* directly exhausts them and leads to premature death, while balance endows the body with the *de*-circulation necessary to transform it in terms of genuineness. The figure achieving this transformation, the Genuine Human, is in almost every respect identical to the Sage described in other early Daoist writings, and this new designation follows from the *Zhuangzi*'s newly configured understanding of the body.

THE *HUAINANZI* DESCRIBES THE CORRELATIVE BODY

The *Huainanzi* represents both the culmination of the early Daoist discourse and the transitional bridge that most substantially links early Daoism with later, institutionalized Daoism. Virtually every major topic treated in earlier Daoist writings is given a sustained discussion within each of the *Huainanzi*'s twenty-one separate essays that are built on pervasive and wholesale quotations from those writings. A huge amount of the earlier writings take episodic, dialogic, or aphoristic form, but the essays of the *Huainanzi* are extremely polished and developed, and are structured by complex internal progressions focused on the general theme named by the title of each chapter. The text's complex thinking and argument advance by exploring the many notions and themes raised in its quotations of the earlier writings without departing from their spirit, and thus the *Huainanzi* can be said to culminate the early Daoist discourse. Its essays are not commentaries on the earlier writings; they are independent works that weave their themes, terms, and images together into a tightly seamed and comprehensive vision of Daoist metaphysics.

The text's extensive use of antecedent materials has tended to lead modern scholars to classify the *Huainanzi* as either an eclectic or syncretic work, and to deny its place in early Daoist discourse. However, these modern classifications should not obscure the fact that the *Huainanzi* not only does belong to that discursive tradition, but it also stands as its last and possibly greatest representative. As an example of the *Huainanzi*'s own claim to membership in early Daoist discourse, we can quickly note the number and kinds of sources that it quotes. According to Charles Le Blanc, there are altogether 842 quotations from earlier works.[74] Of these, 439 are taken from works that Le Blanc identifies as Daoist, while the rest are taken from non-Daoist sources; this number is more than half of the total number. Of these, again, 99 are taken from the *Laozi*, which is remarkable, given the brevity of that text, and 269 are taken from the *Zhuangzi*, more than any other text. Le Blanc also identifies 190 quotations from the *Lüshi Chunqiu*, but he regards that text as belonging to what he calls the "syncretist" tradition without, however, paying due attention to the fact that it contains many notions and themes directly embedded in the early Daoist discourse. Counting the quotations from the *Lüshi Chunqiu* that could be identified as Daoist, one would substantially increase the total number of quotations from Daoist sources.

This is only a formal standard for identifying the *Huainanzi* as a text of early Daoism; in fact, its assumption of the complex of notions and themes constituting early Daoist discourse represents an internal criterion that is as least as persuasive as enumerating the formal number of quotations. Indeed, its exploration of the body goes even deeper than anything that is found in the earlier writings. In many places, the *Huainanzi* pursues its vision of the body by relying on the specific vocabulary taken from the earlier writings; in other places, it *Huainanzi* pursues its vision of the body by applying the technical vocabulary of Five Phase 五行 (*wuxing*) correlative thought. The themes centering on the body play a particularly central role in three of the text's 21 chapters, namely *Huainanzi* 1, 2, and 7.

Huainanzi 1 begins its discussion of the body with an application of two of the terms developed by the *Zhuangzi*, Heaven and the Human, in order to make an immediate and initial distinction between the foundational body and the constructed self: "A person at birth is tranquil, for this is one's nature from Heaven. When that person is stimulated and acts, this is the movement of nature. . . . Thus, those who have attained the Dao do not exchange what is of Heaven for what is of the Human; outside they accompany the changes of things, but inside they do not lose their nature."[75]

In this passage, the *Huainanzi* introduces a distinction between the foundational body as Heaven and the constructed self as the Human, and this will continue to play a central part in its later discussions. The following passage presents a concise description of these terms in relation to the body's physical, fleshly, and manifest nature.

> What I call Heaven is pure quintessence, elemental simplicity, direct substance, pristine radiance, and what has not yet begun to be mixed and blended. What I call the Human refers to a variety of differences; there is crooked cleverness and artful deception that the common person of the world employs for social intercourse. . . . Those who follow what is of Heaven roam in company with the Dao, while those who follow what is of the Human associate with the vulgar.[76]

A further passage in *Huainanzi* 1 also analyzes the foundational body by relying on a set of terms first applied to the body in a systematic way in the *Laozi*, namely, the Dao, *de*, *yin-yang*, and *qi*. It discusses the inner bodily components and the threats to the body that will accrue from not attending to their proper care. These opening phrases are immediately followed by a consideration of the "arrivals" 至 (*zhi*) that refer to the attainment of the Genuine. The *Huainanzi* discusses this in terms of the "Five Arrivals" quoted almost verbatim from *Zhuangzi* 15 (and discussed in the previous section of this chapter). The final lines of this passage bring the entire discussion of "arrivals" back to the physical, fleshly, and manifest body.

> Exuberance and rage are perversions of the Dao; grief and sorrow are losses of the *de*; attraction and aversion are excesses of the heart; addiction and desire are burdens of identity. In humans, great rage shatters the *yin*, and great exuberance destroys the *yang*; shallow *qi* causes muteness, and fear and terror cause insanity; grief and sorrow amass resentments, and the consequence is an accumulation of illnesses; attraction and aversion get tangled up more and more, and the consequence is that excess follows upon excess. Thus, with a heart exempt from grief and elation, he achieves arrived *de*; freely circulating yet unaltered, he realizes arrived tranquility; being relieved of addiction and desire, he realizes arrived emptiness; without attraction and aversion, he realizes arrived equilibrium; not dissipating within other beings, he realizes arrived purity. Being capable of these five things, he is thus able to communicate with the radiance of the spirits. Having communication with the radiance of the spirits, he becomes the master of the inside. Because he uses the center to rule over the outside, the hundred affairs do not go wrong; having obtained it in the center, he is able to reap the fruits of the outside. When the five viscera in the center are pacified, thought and reflection are in equilibrium, the strength of the sinews develops power, the ears are acute and the eyes are discerning, and the *shen* is unob-

> structed and open without confusion. (His body) is firm, solid, and invulnerable; nothing is in excess and nothing is unattainable.[77]

Discussing Heaven and the Human in relation to the inner components of the foundational body, *Huainanzi* 1 applies the notion of achieved genuineness to its description of the physical conditions enjoyed by the cultivated body. The *Huainanzi* as a whole gathers together different ideas from earlier Daoist writings and brings them into a new synthesis, and nowhere is this better demonstrated than in *Huainanzi* 1, which synthesizes the previously mentioned notions from the *Laozi* and the *Zhuangzi* together with the set of notions found in the *Neiye* that the *Laozi* and the *Zhuangzi* do not take up in any consistent fashion. The *Neiye* greatly emphasizes the heart as the center of its soteriological concerns, and also gives particularly prominent positions to the *jing* and to the *shen*, not as well evidenced in the *Laozi* and the *Zhuangzi*. As Harold Roth argues, since the *Huainanzi* consciously takes up these notions inherited from the *Neiye*, the *Neiye* played more of a decisive influence on the *Huainanzi* than on the *Laozi* or the *Zhuangzi*.[78] But more than taking up these notions, the *Huainanzi* integrates them with the set of themes from the *Laozi* and the *Zhuangzi* in a developed presentation that is founded on a grand synthesis of the central vision of the body identified with early Daoist discourse. This synthesis is evident in the following passage, also taken from *Huainanzi* 1, and accords with the tendency found in the *Huainanzi* as a whole that gives a central place to the *shen*.

> The body is the shelter of life; the *qi* is the abundance of life; the *shen* is director of life. When one loses its place, the other two also become damaged. Thus the Sage makes it possible for all people to occupy their proper place and safeguards their proper function so that they do not interfere with each other. Thus if the body does not reside in its proper place of stability, it perishes; if the *qi* is depleted by the absence of what replenishes it, it seeps away; if the *shen* is exercised thoughtlessly, then it obscures; these three things must be safeguarded with extreme care.[79]

Huainanzi 2 radically extends the early Daoist notions of the foundational body by pursuing a historical analysis of the disruption of the body's original constitution, which is somewhat different from the analyses provided by the *Laozi* and the *Zhuangzi*. The historical vision applied in the documentation of this disruption comes as a synthesis of the major ideas informing those earlier analyses. In other words, *Huainanzi* 2 explains the disruption of the body's original constitution as the historical consequence of the breakdown of the original harmony of the world that is inexorably driven to the present state of depravity by the historical facts of real people and their real actions. This appears, for example, in the text's inclusion of dozens of named individuals cited in the historical records and its mention of their decisive actions that have influenced, for better or worse, the present state of human bodies. The primary hermeneutic applied to these discussions is a simple dichotomy that distinguishes the body that is closed off to the world, and thus able to cultivate the internal energies, from the body that fails to maintain its internal integrity. In many ways, this chapter is a direct condemnation of the Confucian programs of self-cultivation, because *Huainanzi* 2 constantly associates the body that fails to maintain its integrity with the Confucian body. The ideal body valorized by the Confucian discourse is, according to early Daoist views, one that has thoroughly displaced its natural constitution, and in its stead has replaced the components of the natural constitution with the artificial adornments that are meaningful only in relation to the social world of custom. In the following passage, *Huainanzi* 2 compares the difference between the foundational body and the constructed self to the difference between a tree growing naturally and the ritual utensils made from its wood.

> When a tree as thick as a hundred arm lengths is chopped down to make a ritual vessel, carved with the engraver's knife, mixed with greens and with yellows painted onto it, and ornamented with floral gems and brilliant stones in the shapes of dragons, snakes, tigers and leopards, then what was curvaceous has become patterned and sophisticated, ultimately only to be tossed into a ditch. If to be whole versus being carved into a ritual vessel means to end up in a ditch, then the difference between ugliness and beauty is great indeed; moreover, the fact is that it has thoroughly lost its nature of being a tree. Therefore, one whose spirit has dispersed speaks flowery words of decorum, and one whose *de* has dissipated carries out artificial actions. At the center, the purest *jing* has perished, and on the outside, they cannot avoid using their body to serve things in their words and actions. Now then, one who hastily and with impunity carries out intentional acts, engaged in the quest for *jing* on the outside, will find that the *jing* has become blocked and exhausted while his acts will have no end in sight; this results in a confused heart and a muddled *shen*, disrupting and perplexing the origins of the individual. What he seeks to preserve is not stable; on the outside he indulges in the customs of the world, causing his judgments to be erroneous and faulty, while on the inside his originally clear illumination is sullied. When this happens he comes to the end of his life completely bewildered, never for a moment having enjoyed peace and contentment.[80]

This passage presents an understanding of how the body's original constitution becomes disrupted and depleted, and it uses images that call to mind the ritualized environment surrounding early Confucian discourse. Another passage from the same chapter presents a more rigorously historical interpretation of the breakdown of the original constitution, identifying each stage of gradual disruption by reference to the different historical ages, beginning from the age of Fuxi. Writings that depict the progression of the historical ages in terms of an ever-deteriorating loss of harmony in the world play a major role in early Daoist discourse, but they do not directly link this progression to the breakdown of the body's original constitution. This passage thus presents a history of the body, or at least a history of bodily illness and discomfort, and aligns it with the progressive history of human civilization. This history as such does not in any significant way deviate from the various histories provided by other early Daoist writings; the *Huainanzi* simply inserts the body into it, and uses the historical ages as a kind of grid to chart the deterioration of the harmony of the foundational body. The final lines from this long passage contrast the early Daoist teachings of the Sage with Confucian teachings, and points out the damage done to the body by pursuing the latter.

> Therefore, the teaching of the Sage is to have one return one's nature to its own origin in order to send the heart roaming in the void. The teaching of the enlightened ones is to have one's nature penetrate into the vast distance in order to awaken to silence and stillness. Now the teachings of the world are not like this, but rather say to pull out vitality and enchain nature, so that on the inside the five organs are worried, and on the outside the senses are overworked. This brings about knots in the stomach due to anxiety, knotting up the strands of things and involving them one to the other. This generates humaneness and righteousness, rites and music, which had until then remained unnecessary. This causes a sudden explosion of knowledge to be taught to the world, becoming the only way to make a name for oneself amidst human society. I find this shameful and will not participate in it.[81]

The passage continues by placing a tremendous blame for the present and depraved state of the body on the shoulders of the Confucians and Mohists: "The disciples of Confucius and Mozi all used the ways of humaneness and righteousness to teach the world, and thus were unable to avoid exhaustion. If their own bodies were incapable of practicing what it is they taught, then how could others practice their teachings?"[82]

As these passages make clear, the most poignant criticism brought to bear against the Confucians is that they do not attend to their foundational body but, rather, use up their store of inner energies by directing their center of attention outside and into the world. This is already extremely harmful to their foundational constitution because they allow those energies to seep out of their bodies and rapidly deplete; what is worse, they actively expend these energies in their programs of bodily discipline that concentrate on the outside comportment of the body in accord with the standards of ritual etiquette and moral deliberation. Expending one's bodily energy in countless acts of kowtowing, blocking the free display of bodily expression by restraining one's natural movements, whenever one is in the company of superiors, equals, or inferiors (in other words, all those situations in which bodily comportment must follow precisely detailed rules), even expending the body's inner energy in the constant control of one's facial expressions: these are the most damaging kinds of activities for the maintenance of the foundational body. The passage even appears to describe a condition eerily similar to ulcers. The *Huainanzi* tends to give special attention to these practices of the Confucians, because they represent the exactly opposite approach to the body taken by the *Huainanzi*. *Huainanzi* 7 gives some very interesting proofs for these criticisms of the Confucian tradition.

> Now as for those Confucians, they do not go to the root of their desires but only attempt to restrict what it is that they desire; they do not go to the origin of their joys but only attempt to block up what it is that gives them joy. This is like keeping the sources of rivers open while trying to cut them off with the hand. . . . Yan Hui, Ji Lu, Zi Xia, and Ran Boniu were the most accomplished students of Confucius. Yan Hui died early, Ji Lu was dismembered and pickled in Wei, Zi Xia was blinded, and Ran Boniu contracted leprosy. All of them did damage to their proper identity, cast off their inner nature, and never attained harmony. . . . In fact their hearts were plunged into mortal depressions, their bodies and internal natures were overworked and exhausted and were incapable of strengthening themselves; therefore none of them were able to last out their Heaven-ordained years.[83]

This passage illustrates early Daoist notions of the early death that will inevitably occur unless a person takes the proper measures necessary to seal up the body and return one's attention to the cultivation of the inner energies, the inherent components of the foundational body. Although this passage appears toward the end of *Huainanzi* 7, it can serve as a proper introduction to the materials that make up the major themes of that chapter. The opening section of *Huainanzi* 7 (discussed in chapter three) depicts the very earliest cosmogonic beginnings from the pristine Dao and the generation of *yin-yang*, Heaven and Earth, and human beings. The section immediately following it gives an analysis of the inner components of the foundational body by stating that these are what the Sage cultivates, and it warns against the dangers of not attending to them.

> For this reason the Sage exemplifies Heaven and follows the course of his natural dispositions: conventions do not constrain him, and the Human does not seduce him. He considers Heaven his father and Earth his mother, *yin-yang* as his principle, and the four seasons as his regulator. Heaven through purity is tranquil, Earth through repose is settled; to displace this is to die, to exemplify this is to live. Tranquility and silence is the stability of the illuminated *shen*; emptiness

> and non-being is where the Dao resides. For this reason, someone who pursues these on the outside displaces them on the inside; one who preserves these on the inside displaces them on the outside. This is as the root is to the branches—yanking them from the root, there is not a single one of the thousand branches, nor a single one of the ten thousand leaves, that does not react. The *jing* and the *shen* are received from Heaven, the form and the body are endowed by Earth. Thus it is said: "The One gives birth to the Two, the Two gives birth to the Three, and the Three gives birth to the ten thousand things."[84]

This passage gives a succinct account of the foundational body: it is born from Heaven and Earth as the parents, with the *jing* and *shen* received from Heaven and the form and the body received from Earth; it is guided by the workings of *yin-yang*; and it is regulated in accordance with the movements of the seasons. Following this, *Huainanzi* 7 applies a set of notions to describe the foundational body that is already familiar from earlier Daoist writings: tranquility, silence, emptiness, and nonbeing. Altogether, these components make up the foundational body, and they must be held together without loss or depletion for them to serve as a continuous source for a longevous body. Displacing them causes physical agitation and ruins the natural harmony of the foundational body. This passage assertively identifies its own depiction of the foundational body with the cosmological environment of *Laozi* 42, thus underscoring the cosmological origins of the foundational body.

The later sections of *Huainanzi* 7, however, open very new directions for the examination and interpretation of the body that at most only existed in partial form in the earlier writings, and these sections of the chapter lay the most important framework for the textual transition from early Daoism to later Daoism. For apparently the first time in early Daoist discourse, *Huainanzi* 7 applies the system of Five Phase correlative thought to the foundational body, and discusses the microcosm and macrocosm relation of the body to the world that is an intimate part of it. Five Phase thought is negligible in the early Daoist writings that predate the *Huainanzi*, and the origins of Five Phase thought is often attributed to a certain Zou Yan (ca. 305–240 BCE) during the later periods of the Warring States. He is not a Daoist, and the earliest applications of Five Phase thought appears to have been directed to analyses of the succession of political dynasties.[85] After a certain period when this thought had begun to circulate throughout early China, it was adapted to discussions of the human body primarily in the early medical literature as the general theory lying behind the diagnosis and treatment of illness, as evidenced primarily in the *Neijing* writings.[86] Looking backward, the *Huainanzi*'s adoption of Five Phase thought, applied in the text's examination of the foundational body, represents the only clearly documented instance of its inclusion within early Daoist writings. Looking forward, Five Phase thought came to permeate thoroughly the writings of later Daoism after the time of its textual application to the body in the *Huainanzi*. For these reasons, *Huainanzi* 7 is a remarkable chapter because it textually links the major themes cohering around the early Daoist visions of the foundational body with the primary interpretative framework with which later Daoism presented its own visions of the body.

The initial sections of *Huainanzi* 7 depict the foundational body in terms already established in early Daoist discourse before launching into its original depictions in terms of Five Phase thought. The two sets of terms are not exclusive, and the text weaves them together in a seamless whole as it explores the relation between the body as microcosm and world as macrocosm. The first portions of this section present the different stages of the development of the embryo during each of the ten months of its gestation:

> Thus it is said, in the first month there is the embryo; in the second month there is swelling; in the third month there is the fetus; in the fourth month there is flesh; in the fifth month there are sinews; in the sixth month there are bones; in the seventh month the organism is completed; in the eighth month there is movement; in the ninth month it is restless; in the tenth month there is birth. The five viscera form as the body develops.[87]

This passage naturalizes the gestation period in anatomical terms while at the same time, in designating this period through a ten-month progression, it sets the stage for the macrocosmic correlations immediately following.

Huainanzi 7 goes beyond the basic parameters structuring the earlier Daoist discourse by discussing in detail the internal anatomy of the human organism in terms of the five viscera 五藏 (*wuzang*). These are the lungs, the kidneys, the gall bladder, the liver, and the spleen. Even without the assistance of Five Phase terminology, earlier Daoist writings never concentrated their attention on these viscera or any of the other internal organs in a systematically anatomical way, preferring instead to attend more exclusively to the inner components in less anatomical form. Beginning with the stages of the ten-month gestation, in which the embryo comes to life by congealing and incorporating the cosmic energies, *Huainanzi* 7 concentrates its discussion on the internal anatomical structure of the foundational body, in which the viscera and other organs are seen as the congealed orbs of the cosmic energies in manifest form, around which the body forms: "The five viscera form as the body develops."

The discussion of the five viscera that follows provides an account of the foundational body in terms of the Five Phases, together with the correlations that demonstrate their identity with the macrocosm. These correlations further extend the limits of what the earlier Daoist writings had taken as their focus, but this discussion in *Huainanzi* 7 does not depart from the specific motive informing the discussions of the foundational body in the earlier writings: to provide an explicit account of the body's possibilities for embodying the pristine Dao. The terminology of Five Phase thought simply was not available to the earlier writers. The possibilities that Five Phase terminology opened up for pursuing the discussion of the foundational body were so powerful and so in keeping with the specific goals of the earlier Daoist writings, as the *Huainanzi* goes to great lengths to demonstrate, that it would not be the least bit surprising to see it applied in the earlier texts, if it had been available to the writers. The adoption of this terminology in the *Huainanzi* is explained more correctly by seeing it as the timely recipient of the newly developed Five Phase terminology rather than by claiming that it radically departs from the spirit of earlier Daoist discourse.

The rest of the discussion of the body in *Huainanzi* 7 is devoted to furthering this understanding of the foundational body. Moreover, the text maintains the central position given to it within the total vision pertaining to early Daoist cosmology and soteriology. It discusses the necessity of sealing up the vital energies of the body in order to cultivate them directly, as well as the soteriological union with the Dao within the physical space of the foundational body. In the following passage, I provide in brackets the explanatory comments supplied by Gao Yu at the beginning of the third century CE; these comments bring out the underlying correspondences that the original text does not give explicitly, but that may have been recognizable to the original author.

> Now the lungs regulate the eyes. [Lungs image the red sparrow; the red sparrow is fire; fire illuminates the outside world, thus it regulates the eyes.] The kidneys regulate the nose. [Kidneys image the tortoise; the tortoise is water; water flows through waterways, and the *qi* circulates through the nose, therefore they regulate the nose.] The gall bladder regulates the mouth. [Gall bladder is courage and bursts through its housing, therefore it governs the mouth.] The liver governs the ears. [Liver is metal; metal is illuminated from within, therefore it governs the ears.][88]

The commentary points out that although the spleen, the fifth of the five viscera, is not included in this passage (possibly due to a textual corruption), it is included in a similar passage discussing the five viscera in *Huainanzi* 3. There the text states that the spleen governs the tongue; in the next section of this chapter, the spleen is discussed as one of the five viscera. It may also be possible that the writer had not yet formalized the list.

Huainanzi 7 presents a set of correlations that associate the separate aspects of the macrocosmic realm with the microcosmic body. These correlations, too, do not radically depart from the spirit of the earlier Daoist writings. Specifically, in the *Zhuangzi*, the total identity of the foundational body was not limited to the spatial extent of the four limbs, but was described as one with the ten thousand things and Heaven and Earth. These are the ideas that inform images such as those found in *Zhuangzi* 2: "Heaven and Earth were born together with me; I and the ten thousand things are one." The discussions of the relationship between the macrocosm and microcosm in *Huainanzi* 7 take these notions of the total identity of the body at face value, but by adopting the model of the macrocosm and microcosm (an integral component of Five Phase thought), the text is able to designate these relations in new ways. I provide Gao Yu's comments in brackets in this quotation for the reasons previously stated.

> Thus the head is round in the image of Heaven, the feet are square in the image of Earth. Heaven has Four Seasons, Five Phases, Nine Divisions, and Three hundred sixty-six days. Humans also have Four Limbs, Five Viscera, Nine Openings, and Three hundred sixty-six joints. In Heaven there is Wind and Rain, and Cold and Heat. Humans also have Taking and Giving, Joy and Anger. Thus there is a correspondence of the gall bladder with the clouds [gall is metal; from metal and stones clouds emerge, thus it corresponds to clouds], the lungs with the *qi* [lungs are fire; thus they correspond to the *qi*], the liver with wind [liver is wood; wood causes wind to be born, thus it corresponds to wind], the kidneys with rain [kidneys are water; because of water there is rain], and the spleen with thunder. The Human forms the third term with Heaven and Earth, and the heart acts as master [heart is soil, and thus it is the master of the other four phases].[89]

The reference to the three-part world consisting of Heaven, Earth, and the Human serves to underscore the specific motives behind the *Huainanzi*'s adoption and application of the model of the macrocosm and the microcosm. The passage explores the ultimate consequences of the total identification of the body with the world by relying on the technical terminology of Five Phase thought and of the microcosm and macrocosm.

The discussions of the anatomical structure of the foundational body in *Huainanzi* 7 present a magnificent synthesis of the terms, images, and themes informing early Daoist writings on the body, on the one hand, and the terminology of Five Phase thought, on the other. Either as cause or consequence of this synthesis, the text describes the inner components of the foundational body in terms of their manifest localizations as internal organs; in other words, the internal organs are concrete, anatomical manifestations of the inner components regulated by the rhythms of the Five Phases. Earlier Daoist writings do not discuss the concretized presence of these components in terms of organs, but only abstractly: *yin-yang* represents the physical and energetic flows of the body, and *de* describes the body's powerful circulation; *jing* is the fluid matter of the body, *qi* is the breath that coarses throughout the body, and *shen* is mind or spirit. These components were never localized with any particularization within the anatomical structure of the body in early Daoist discourse before the *Huainanzi*; its description of these inner components localized in each of the viscera marked a further extension of the limits defining early Daoist writings. It is within the scope of these discussions that *Huainanzi* 7 introduces its soteriological considerations concerning the embodiment of the pristine Dao within the foundational body: if the body is sealed off, then the internal components can be cultivated and the body will transform, thus opening the way for the direct embodiment of the Dao in the body.

> How can the ears and eyes of human beings be able to exert themselves through the course of time without respite? How can the *jing* and the *shen* wildly gallop on without repose? Thus the blood and the *qi* are the florescence of the flourishing of human being, while the five viscera are the very *jing* of human being. If the blood and *qi* are able to concentrate within the Five Viscera instead of being dissipated outside them, then the breast and the belly fills up and passion and desire minimize. When the breast and the belly fill up and passion and desire minimize, then the ears and the eyes are clear and hearing and sight are far-reaching; this is called illumination. If the Five Viscera are placed in dependence upon the heart and do not stray, then refined intention prevails and one's conduct does not deviate. When refined intention prevails and one's conduct does not deviate, then the *jing* and the *shen* are abundant and the *qi* does not dissipate. When the *jing* and the *shen* are abundant and the *qi* does not dissipate, then everything is ordered; when everything is ordered then there is equilibrium; when there is equilibrium, then there is passage; when there is passage, then there is sacrality; when there is sacrality, then there is a vision in which nothing is not seen, a hearing in which nothing is not heard, and an acting in which nothing is not accomplished. Thereby sorrow and grief find no entrance, and pernicious *qi* is unable to invade. Thus busying oneself with searching [for the Dao] throughout the Four Seas will not help one to find it, nor will jealously guarding it within the interior of the body help one to see it.[90]

The passage that immediately follows presents the extreme counter-example of the body that is not sealed off and goes on to describe the terrible consequences that will accrue because of this.

> When the ears and eyes are seduced by the pleasures of sounds and sights, then the Five Viscera are shaken, waver and lose their stability. When the Five Viscera are shaken, waver and lose their stability, then the blood and *qi* are agitated, overflow, and find no repose. When the blood and *qi* are agitated, overflow, and find no repose, then the *jing* and the *shen* wildly gallop off outside and are not maintained within. When the *jing* and the *shen* wildly gallop off outside and are not maintained within, then calamity and blessing arrive, and although they are as big as hills and mountains, there is no way of knowing from where they come.[91]

These discussions of the foundational body are ultimately resolved in the depictions of the Genuine Person as one who has transformed the body so that it continuously embodies the Dao within. Because of this embodiment, the Genuine Person perpetually to rejuvenates the body's inner energies, which are identified with the energies of the world itself and allow for the total identity of the body with the world as a whole.

> For this reason, in his roamings, the Genuine Person, expelling and sucking, exhales and inhales; spitting out the old he renews the inside. With bear-steps and bird-stretches, duck-ablutions and monkey-jumps, owl-stares and tiger-gazes, he nurses his body in order to keep his heart from slipping away. He makes the *shen* flowing and pure without losing its fullness, and in the alternations of day and night there is nothing to damage it; he thus acts as the spring for all things. This then is to gather everything within and give birth to the seasons within the heart. This person has a purified body and the heart has no decline; he has a completed lodging and the *jing* never diminishes.[92]

Early Daoist discourse is fundamentally anchored by the central position given to the physical, fleshly, and manifest body, and the *Huainanzi*'s vision of the body inherits and radically extends various notions of the body from the *Neiye*, the *Laozi*, and the *Zhuangzi*. The central themes of those writings focus primarily on the inner components of the foundational body in contrast to the constructed self that requires dangerously high levels of energy consumption. The continuous maintenance of the constructed self ultimately results in the exhaustion of the natural energies of the foundational body, and early death. Those writings consistently exhort one to seal off the body in order to expel all that belongs to the constructed self, and directly cultivate the inner components of the foundational body. Doing this causes a radical transformation of the body that leads to a realization of its total identity with the world and the cosmos as a whole. The figure that is commonly said to possess this kind of transformed body is named as the Sage or the Genuine Person; he is physically united with the Dao and makes the Dao present in the world by embodying it in this very body in this very world. The soteriological consequence of this is that the world attains a second-order harmony, and the body of the Genuine Person attains longevity. The longevity attributed to the Genuine Person is not the lasting quality of the body as we know it; rather, the transformed body of the Genuine Person is able to change and transform at will, combining with the different forms of nature while still maintaining possession of the integrity of its inner components. *Huainanzi* 7 describes the transformations of the Genuine Person.

> The inner nature of the Genuine Person is united to the Dao. He exists but appears non-existent; he is actual but appears empty. . . .

He regards death and life as transformations of the One, and all things as coming from the recipes of the One. He shares the *jing* of the root of great purity, and roams through the edges of the regions of the indistinct. He possesses the *jing* but does not manipulate it; he possesses the *shen* but does not labor it; he merges with the simplicity of the great non-separation and takes his place in the center of supreme purity. For this reason his sleep is without dream, his wisdom is without harbor; his *po*-souls do not submerge, his *hun*-souls do not fly off. He moves back and forth between beginnings and ends, and nobody can distinguish his emerging or parting. He gently closes his eyes in the house of the great night, and awakens in the home of the shining clarity; he rests in the recesses of the untwisted, and goes roaming in the wilds of the unformed and undifferentiated. He resides in the featureless, and he makes his home in the regions where nobody goes.[93]

Chapter Six

Early Daoism and Modernity

Throughout the composition of this work, I constantly had to ask myself what possible value a historical study of a relatively small-scale early Chinese tradition of discourse might have for readers in a world such as the one we all live in today. This is not to give the impression that I have any expectations whatsoever of making the *New York Times* best-seller list, only to point out that we all need to locate our motivations for the projects we undertake and find some solace for the decisions we make that eat up the years of our allotted lives. When people have asked me about my work, they come to the immediate conclusion that I am doing philosophy of, if not an obscure kind, then at least of a generic Eastern sort whose reputation took some beating with the homogenization of our modern Western society beginning with the Reagan years and the computerization of our professional and personal lives. "Eastern philosophy" sometimes raises nostalgic brows from those of my generation, with memories of a young adult fascination with this material that was left behind when the glow of college curiosity wore off several years into the workaday world. The issues that motivated me to pursue and complete this work, however, were more religious than philosophical.

Recently surfing through the cable news channels one evening, I found a debate about religious freedom, an issue, not unexpectedly, argued from a decidedly political perspective concerning the flag or the Pledge of Allegiance or something like that. One person mentioned that in America today, "We have Christians, Jews, Muslims, and, umm, oh yeah some Hindus too. Therefore we need to protect religious freedom in the name of our founding fathers." I think I am too jaded to continue to be appalled by such remarks. It is not simply the question of representation in the modern media and respect for others in a diverse society that irked me in that spot about religion; it's the question of just what it is that we have in mind when we talk about religion; what meaning do we give to religion as it exists in the world today?

I have described my book about early Daoism as a work of religious history, and this typically confuses people. It is hard to understand how something as devoid of institutional foundation as early Daoism appears to be could be seen as a religion. My Western friends find no social organization, set of doctrines and dogma, or postlife continuation, and my Chinese friends are as perplexed; didn't the CCP define religion according to its five Characteristics: complex, mass-based, long lasting, has implications for ethnic nationalities and with the other nations?[1] (Early Daoism as a religion fails under numbers 1, 2, 4, and 5.) In the West, definitions of religion are not so clear-cut, but we all know that Christianity, Judaism, and Islam qualify, and we grudgingly allow that status to Hinduism and Buddhism. This is despite the fact that typical college textbooks present a far higher number of world religions, but it seems that they are stretching by about the sixth entry. Our reluctance to leave behind the institutional factor in our understanding of religion is understandable, after more than a thousand years worth of effort to establish institutionalized Christianity as the one true religion.

Philosophers throughout history were able to more or less stay clear of charges of heresy, yet even Descartes admits the undeniability of God standing at the origin of his reductions. Some centuries earlier, philosophy was not all that different from religion. For the ancient Greeks, philosophy was not situated in academic institutions and contained within disciplinary walls, though Plato did have an academy. Philosophy was a way of life directed to the realization of wisdom and heightened spiritual states of awareness; but at the dawn of philosophy as we know it stand shady cults and shadier practices. It was the prophetess Diotima, speaking within the shadowy confines of her temple dedicated to the gods, or so we imagine, who saw the ladder to wisdom rooted in *eros*, erotic attraction, a wisdom that was, to Socrates at least, attainable through the moral renunciations newly introduced into the pederastic games of the time.[2] It wasn't until around the sixteenth century, explains Pierre Hadot, that we see philosophy separate itself once and for all from such "spiritual practices" to take its present place in the lecture hall.[3]

The social reality of early Greek philosophy radically changed in the Roman world. More and more minds were drawn to the possibilities offered by theology, with its social organization plotted around saints and bishops, and the quest for wisdom slowly gave way to lives lived for salvation. The ancient relation between religion and philosophy was not of a kind familiar to us today. Here, philosophy lives in the academy, while religion appears pervasive in all other places including, in many instances, the academy. Religion is pervasive, and the great majority of us have one.

Clearly it matters what we mean when we say religion, particularly if we mean something more inclusive than Christianity, Judaism, Islam, and, um, Hinduism too. It seems to me that the notion itself ought to include something essential that is not often recognized, namely, its function for sense. Religion appears as a primary component of that which allows us to order our experience of existence in the world. Among other things, religion is a discourse, actually a vast forum for countless discourses. These discourses have a common function in that they provide us with the material for creating or discovering, inheriting or maintaining, the different meanings with which we invest the world. Religion, in its capacity for ordering the sense of the world, is fundamentally violent in that it cuts the world up into lots of pieces: high-low, sacred-profane, good-evil. Our understanding of the world seems eminently grounded in the imposition of religious distinctions, about space (from mountains to nations), time (in either its linear or cyclical form), shapes (including crosses and gourds), and bodies (saved or damned, immortal or empty). This appears as a practice found both at the beginning of recorded history, as well as in the present age.

Another typical element of religious discourse is that *quelque chose d'autre*, that something beyond, not necessarily in a transcendent sense of God or Absolutes, but in the sense of that unknown origin on which all human sense and civilization is built, whether we name it God, Dao, Brahma, turtles, or reeds. That original unknown, that *quelque chose d'autre*, somehow continues to exert its presence and has a privileged position at the center of religious thought and activity; it is something to which human beings continue to respond. The modern existentialist movement represents an odd case, in that it inquires into the conditions of meaning and sense while deliberately resisting recourse to that unknown something: "existence precedes essence."[4] We are thrown into the world, and the world is devoid of any meaning, any significance whatsoever, so they say. Maybe the greatest question of all is, as Heidegger claimed, the question of why there is something rather than nothing: why do we feel the need to invest the world with meaning at all?[5] Why does religion, that greatest of all investors of meaning, have to be? Or, to ask the historical question, why do all known human cultures subsist on a bedrock of meanings both inherited and adaptable? What would it mean to find a culture without its own meanings of the experience of the world, without its own religion? Western explorers thought they had found some from time to time; the Jesuits for a long time hailed imperial China as one such example. Communists in the last century also thought they had created other such cultures. America, this most developed of all possible nations, certainly is not one such culture.

Religion provides the raw material for the creation and maintenance of meaning. To varying degrees depending on time and place religion is open to debate and adaptation, and its applications to the meaning of the war on terror mark but one relatively insipid example of this. The meanings provided by religion are deeper, more fundamental to our makeup, and more worthy of our consideration (so we should hope) than current fad, even if our current fads might spell the end of our world, as we know it. Outside of the question of returns, of true and false or right and wrong religion, the meanings it supplies are the pieces that we use to construct our spaces and find ourselves at home in the world.

History plays a central role in the maintenance and transmission of the meanings of religion. Religious meaning, in large part inherited from the past, adapts in step with historical change; in this sense, science is her great grandchild, who has only very recently attempted to wrench himself free to stand alone. That *quelque chose d'autre*, the original unknown, remains ever elusive even to the smartest physicists. Can there be specifically scientific grounds to argue for war and revolution? We still use the ancient texts to justify, moralize, or interpret modern conditions and situations: the *Mencius* speaks to the Confucian sense of the right to revolution, while the Bible and Koran speak to the Jewish, Christian, and Muslim sense of vengeance, justice, and acquittal. Distancing myself from these blurred lines of religion and politics, I want to ask: do examinations of the ancient texts provide us with any ideas that we ought to take seriously?

In this, I think we can rule out the possibility of locating hard truths. Although it is interesting to find certain modern readings of the Buddhist sutras that explain relativity, or those that recommend the study of the Om kibbutz for decoding DNA, world scriptures only rarely serve as sources for solid history. Putting aside for the moment the historical relevance of such truths as the existence of Heaven, the presence of immortals, or the comforts of nirvana, a different kind of investigation, asking different sorts of questions, seems more appropriate to the inquiry into the contemporary value of the ancient texts. While both scientific empiricism and common perception would argue against those and other such truths, this does not by any means exhaust the limits of the question of the meanings of the sense of the world made available through the various religious discourses.

There are many ways of cutting up the religion pie; one that makes some sense to me is the separation between the polytheistic and monotheistic religions. While many polytheistic traditions celebrate the existence of a high god, from ancient Egyptian religion to modern Siberian shamanism and most others in between, what sets them apart is their attention to less celestial deities, like earth goddesses, recognized by many to populate the earth, whereas the monotheistic religions, while not in every case entirely denying the presence of spiritual powers active on earth, focus worship on one high god. Despite these radical differences, both sorts of religion actively share in the construction and maintenance of the meanings of the experience of the world. They make the world sensible in ways that otherwise, historically speaking, are literally unthinkable. Many religions, in fact, take this aspect of religion as their central teaching, and scholars of ritual tell us that this comes as no surprise to them. One needs merely to take up almost any recent study on Vedic ritual, for example, to witness the metamechanics, or metaphysics, of world maintenance.[6] Bodhidharma is said to make the following speech, which seems appropriate for me to cite here:

> The brush of mind and the consciousness discriminates and draws forms, sounds, smells, tastes and touchables, and, upon looking at them in turn, produces greed, anger, and stupidity. Sometimes it is fascinated and sometimes repelled. Due to the discriminations of thought, mind, and the consciousness, various sorts of karma are in

> turn produced. . . . Some, by the discriminations of their own mind, draw tigers, wolves, lions, poisonous dragons, evil spirits, and the generals of the five paths of rebirth. . . . These things are discriminated by their own minds, but they are then controlled by these things, and so they undergo various sufferings.[7]

Different religions paint their own pictures, though some of these pictures sometimes closely resemble those of other religions; polytheistic brushes seem to paint pictures more akin to other polytheistic religions across time and space than they do to monotheistic pictures, and vice versa. Certain kinds of beings and places also tend to assume positions of importance as valued within each of the two sorts of religion; God, Heaven, and Hell play a central role in the monotheistic religions, while gods, rivers, and mountains are more up front on the polytheistic screen. Regardless of these differences, both express complete views of the world in sensible, more or less coherent, and fundamental ways.

As a religious discourse, early Daoism also presents a sensible, more or less coherent, and fundamental picture of the meanings of the experience of the world. That world is informed, in many important ways, by a view that is Dao-centered; their picture of a Dao-centered world is constructed, maintained, and transmitted through the tradition of discourse to which this work is committed. In examining this discourse, I have striven to understand the world painted by those early writers through certain of my own categories that allow me to make sense of their discourse: cosmogony, cosmology, ontology, and soteriology. While my categories could be challenged as labels inappropriately applied to early China, the bottom line is that they are the best tools available to me in my effort to make sense of that material.

Although this work is a study of one religious discourse among many available to the scholar, it remains first and foremost a book of history. Although we know next to nothing about those who formulated and wrote the contents of this discourse, they were real people who slept, ate, lived, and died. How much of what they wrote can be taken as representative of the commonsense experience of others sharing their time and place, and how much as representative only of their own personal ideas independently conceived, are things that we will most likely never know. We do know, however, that their ideas, images, and aspirations infused Chinese culture in a fundamental way after the unification of the empire in the third century BCE, and the presence of Daoism even today is palpable throughout most of the Far East and is also not negligible elsewhere. Like those of other religions, the discourse of early Daoism should not be limited to issues of right or wrong, true or false; rather, the issue with such discourses is also about the meanings of the experience of the world that we, as participants in any given culture, inherit, create, and maintain.

The discourse of early Daoism, like those the world over, fudges its status as history or eternity. As history, we approach it from the outside, seeking to learn how certain people configured their experience of the world in terms of meaning; as eternity, we listen to them from the inside, bearing witness to their timeless truths capable, or not, of unraveling the mystery of the world and our reasons for being in it. These two approaches, not only to Daoism but to all the religions, have had and continue to have powerful representation, and neither ought to be dismissed out of hand by proponents of one side, since neither approach can, once and for all, be entirely wrenched from the other. This is, indeed, a major component of the reason we count such texts as scriptures and classics to begin with. But the question remains, why should we care about what these discourses have to say to us in the modern age? What difference do they make?

If, as I have tried to argue, it is reasonable to see in religion that capacity to create, transmit, alter, and maintain our meanings of the experience of the world, then, for any one of them to maintain a privileged status among those who hold to it, it must be able to adapt. This necessity to adapt is, at the very least, a simple consequence of participation in history itself. The Bible needs the Vatican for its continued relevance to Catholics just as much as Muslims depend on the fatwa's of the imans; one can go into a Christian church most anywhere in Taiwan and find a statue of Buddha or Confucius to help one's prayers get through. The religious meanings handed down from our ancestors are in need of constant maintenance. I believe that there is a good argument to be made for the benefits of an open mind, and not just for the sake of multicultural dialogue. It pertains to being able to make our meanings of the experience of the world relevant in changing situations. Whether looking to the classics is a sign of cowardice or courage, this is what we do. I don't think there is a tremendous difference if those texts we refer to are the Bible, the Constitution, the Hadith, or the Liji; our ability to make continuing sense of the world depends on the discourses we inherit from the past; as long as we can locate the appropriate discourse from within our cultural experience, we can figure out how best to manage change. But if each culture has its own scriptural traditions on which to fall back, why should we bother with those not belonging to our own?

For no other reason than to become aware of the limits imposed by each religion, inevitable victims, every now and again, of their own possibilities negated by influential times and people who take up and promulgate any set of meanings for the experience of this world, as we know it. Galileo wrought havoc in the placid world of medieval theology, as did Darwin to the Victorians, as did the Muslims to Hindu India and as did Christian imperialism to late imperial China; the list goes on. Religious views of the possibilities of the world do indeed break down, but they are only very rarely buried and destroyed, once and for all. The present world experiences different sorts of tensions, concerning consequences never thinkable before the last century. We are reluctant to admit the limits of our own meanings of the experience of the world, reluctant to admit the limits of our own discourses, even when not doing so results in paradox and pain.

The various discourses of each of the religious traditions of the world have been studied, and will continue to be studied, by people of all sorts. They inspire us to consider the meanings of the experience of the world in ever-deeper ways. Like ritual, engagements with these texts allow us to reconfigure our understandings of the world, providing the material and motivation to reconsider the limits of what we are able to think. Above and beyond the question of true and false, right and wrong, the present work on the religious and historical discourse of early Daoism is simply an indication and invitation to rethink one of the great discourse traditions that has provided a view of the world that is sensible, coherent, and fundamental. Can it not help us think?

Notes

CHAPTER ONE

1. For an interesting study on the relation between abstract concepts and metaphor, see George Lakoff and Mark Johnson, *Philosophy in the Flesh: The Embodied Mind and Its Challenge to Western Thought* (New York: Basic Books, 1999).

2. The actual history of the Jixia Academy and its importance as a center of philosophical thought are issues that are not unchallenged among specialists today. I only mention the Jixia Academy here in order to suggest ways in which the emergence of early Daoism can be generally dated and situated. For arguments favoring the existence of the Jixia Academy and its importance as a center of philosophical thought, see R. P. Peerenboom, *Law and Morality in Ancient China* (Albany: State University of New York Press, 1993), especially pages 224–242; for arguments against its existence, see Nathan Sivin, "The Myth of the Naturalists," in *Medicine, Philosophy and Religion in Ancient China* (Great Britain: Variorum, 1995), pages IV 1–33, and Nathan Sivin and Geoffrey Lloyd, *The Way and the Word* (New Haven: Yale University Press, 2002). For a discussion of the Guodian finds and their relation to early Daoism, see Thomas Michael, "Confucius and Laozi: Two Faces of the Dao," in *Metaphilosophy and Chinese Thought*, eds. Ewing Chinn and Henry Rosemont, Jr. (New York: Global Scholarly Publications, 2005).

3. Herlee Creel, *What Is Taoism? And Other Studies in Chinese Cultural History* (Chicago: University of Chicago Press, 1980), 41.

4. Creel, *What Is Taoism?*, 46.

5. Creel, *What Is Taoism?*, 24.

6. See A. C. Graham, *Disputers of the Tao: Philosophical Argument in Ancient China* (LaSalle: Open Court, 1989), 186–199; 306–311; and "How Much of the *Chuang-tzu* did Chuang-tzu Write?" in *A Companion to Angus C. Graham's Chuang Tzu*, ed. Harold Roth (Hawaii: University of Hawaii Press, 2003), 58–103.

7. For a different set of arguments that come surprisingly close to Graham's, but that have not received the same amount of scholarly attention, see Liu Xiaogan, *Classifying the Zhuangzi Chapters* (Ann Arbor: University of Michigan Press, 1994).

8. Harold Roth, *Original Tao: Inward Training (Nei-yeh) and the Foundations of Taoist Mysticism* (New York: Columbia University Press, 1999), 7.

CHAPTER TWO

1. See, for example, Mircea Eliade, *The Myth of the Eternal Return: Or, Cosmos and History*, tr. Willard Trask (Princeton: Princeton University Press, 1991).

2. See, for example, Jenny So, "Chu Art: Link Between the Old and New," in *Defining Chu: Image and Reality in Ancient China*, ed. Constance Cook and John Major (Hawaii: University of Hawaii Press, 1999), 43; John Major, "Characteristics of Late Chu Religion," in *Defining Chu: Image and Reality in Ancient China*, eds. Constance Cook and John Major (Hawaii: University of Hawaii Press, 1999), 124–125.

3. So, "Chu Art," 43.

4. Norman Girardot, *Myth and Meaning in Early Taoism* (Berkeley: University of California Press, 1983), 203.

5. Mark Lewis, *Sanctioned Violence in Early China* (Albany: State University of New York Press, 1990), 204.

6. Li Ling and Constance Cook, "Appendix: Translation of the Chu Silk Manuscript," in *Defining Chu: Image and Reality in Ancient China*, eds. Constance Cook and John Major (Hawaii: University of Hawaii Press, 1999), 174.

7. *ZZ*, 1:1a–1b.

8. *ZZ*, 1:1b.

9. *ZZ*, 1:4a.

10. Girardot, *Myth and Meaning in Early Taoism*, 203.

11. The paradoxical play of distinctions in spatial and temporal perspectives is furthered developed throughout the course of *Zhuangzi* 1.

12. *XCZ*, 75.

13. *XCZ*, 76.

14. Gerald Swanson, *The Great Treatise: Commentary Tradition to the "Book of Changes"* (Ann Arbor, Michigan: UMI Dissertation Services, 1974), 116–117.

15. *XCZ*, 84.

16. *XCZ*, 75.

17. See Allyn Rickett, *Guanzi: Political, Economic, and Philosophical Essays from Early China, Volume 2* (Princeton: Princeton University Press, 1998), 98.

18. *SD*, 14:1a–1b.

19. The virtually identical process is found in the *Taiyi Sheng Shui*.

20. *SD*, 14:2a.

21. This passage shares strong similarities with the description of the gestation period found in *Huainanzi* 7.

22. *SD*, 14:2b–3a. Text amended following Rickett, *Guanzi*, 105.

23. Note that this text represents a possible source for Zhen Wu 真武 and his association with the snake. Zhen Wu is associated with North, which is water in the later theory of the Five Phases.

24. *HNZ* 7, 7:1a.

25. For a very different view of this, see Michael Puett, *To Become a God: Cosmology, Sacrifice, and Self-Divinization in Early China* (Cambridge: Harvard University

Press, 2002), pp. 1–3. For a brief challenge to his ideas, see Thomas Michael, "Debating the Spirit in Early China" (review article), in *Journal of Religion*, v. 83, (2003), 421–429.

26. See Sarah Allan, *The Way of Water and Sprouts of Virtue* (Albany: State University of New York, 1997) for a more general survey of this textual phenomenon in early China.

27. *LZ* 8, 1:4b.

28. *LZ* 4, 1:4a.

29. Girardot has studied the associations between gourds and the Dao in relation to the theme of chaos and cosmology in *Myth and Meaning in Early Taoism.*

30. *ZZ* 7, 3:17.b–18a.

31. *ZZ* 7, 3:18a.

32. *ZZ* 7, 3:18a–18b.

33. *ZZ* 7, 3:18b.

34. *ZZ* 6, 3:2a.

35. *LiZ*, 2:16b–17a.

36. Cited earlier: "The Dao is empty, yet when you use it, you never need fill it again. Like an abyss! It seems to be the ancestor of the ten thousand things. . . . Submerged! It seems perhaps to exist. We don't know whose child it is; it seems to have preceded Di."

37. Li Ling, "An Archaeological Study of Taiyi 太一 (Grand One) Worship," tr. Don Harper, *Early Medieval China* 2 (1995–1996): 1–39, and Don Harper, "The Nature of Taiyi in the Guodian Manuscript 'Taiyi sheng shui': Abstract Cosmic Principle or Supreme Cosmic Deity?" Paper prepared for the "International Symposium on the Guodian Chu Bamboo Slips and Related Excavated Materials" (December 2000).

38. Li, "An Archaeological Study of Taiyi," 25.

39. Harper, "The Nature of Taiyi in the Guodian Manuscript," 2. For a similar view, see also Puett, *To Become a God.*

40. *TYSS*, 126. Text amended in accordance with Harper, "The Nature of Taiyi in the Guodian Manuscript," 4, reading *bo* 薄 ("join") for *fu* 輔 ("assist") as emended by the editors of the manuscript.

41. For a more detailed study of this mechanism in the *Guodian Laozi*, see *The Guodian Laozi: Proceedings of the International Conference, Dartmouth College, May 1998*, eds. Sarah Allan and Crispin Williams (Berkeley: Society for the Study of Early China and Institute of East Asian Studies, University of California, 2000), 162–170.

42. *LSCQ*, 5:3a–3b.

43. *LZ* 52, 2:10a. The received text is identical to Mawangdui text.

44. *GDLZ*, 126.

45. *MWDLZ*, 266.

46. *LZ* 1, 1:1a.

47. *MWDLZ* 51, 72. The received edition writes: "Dao gives birth to them and *de* nourishes them." Other differences are minor.

48. *LZ* 5, 1:3b.

49. Li, "An Archaeological Study of Taiyi," 1.

50. *LZ* 25, 1:13b–14a.

51. *MWDLZ* 14, 284. The received edition does not have the second "One;" other variations are minor.

52. *DY*, 172–174.

53. *HNZ* 1, 1:1a–1b.

54. *HNZ* 1, 1:10a–10b.

55. Girardot has examined these early Daoist writings that represent the origins in terms of sacs, wombs, and gourds that burst open at the moment of birth; see his *Myth and Meaning in Early Taoism*.

56. *LZ* 21, 1:11b–12a.

57. *ZZ* 7, 3:19a–19b.

58. *LZ* 78, 2:23a.

CHAPTER THREE

1. For more on the journeys of the *wu*, see Rémi Mathieu, *Étude sur la mythologie et l'ethnologie de la Chine ancienne* (Paris: Institut des Hautes Études chinoises, 1983).

2. *ZZ* 23, 9:7a–7b.

3. Here I argue against the imputation of transcendentalism to early Chinese thinkers, especially as formulated in Heiner Roetz, *Confucian Ethics of the Axial Age* (Albany: State University of New York Press, 1993).

4. For more on these ideas, see Edmund Husserl, *The Crisis of European Sciences and Transcendental Phenomenology* (Evanston: Northwestern University Press, 1970).

5. See ch. 2 for my discussion of the first part of this writing.

6. *TYSS*, 126–129.

7. For the early Daoist writings, it was a very small step to substitute the Human for the Sage to stand as the third term in representations of the world, and this is precisely what occurred, as is amply evidenced in the later writings.

8. *LZ* 14, 1:7b.

9. See William Boltz, "Kung Kung and the Flood: Reverse Euhemerization in the *Yao Tian*," *T'oung Pao* 67 (1981): 142, and Deborah Porter, *From Deluge to Discourse: Myth, History, and the Generation of Chinese Fiction* (Albany: State University of New York Press, 1996).

10. *LZ* 21, 1:12a.

11. *LZ* 25, 1:14a.

12. *ZZ* 25, 8:29b–32a.

13. *TTYS*, 129.

14. *LZ* 47, 2:7a–7b.

15. *LZ* 34, 1:19b.

16. *LZ* 17, 1:9b.

17. For more on the placing of deities in the body, see Kristofer Schipper, *The Taoist Body*, tr. Karen Duval (Berkeley: University of California Press, 1993), especially ch. 6; for more on the construction of the altar, see John Lagerwey, *Taoist Ritual in Chinese Society and History* (New York: Macmillan, 1987).

18. Roger Ames, "Introduction," in *Wandering at Ease in the Zhuangzi*, ed. Roger Ames (Albany: State University of New York Press, 1998), 3.

19. *LZ* 23, 1:12b–13b.

20. *LZ* 75, 2:22a.

21. *LZ* 53, 2:10b–11a.

22. *LZ* 32, 1:18a–18b.

23. *LZ* 37, 1:21a.

24. *LZ* 57, 2:12b.

25. *LZ* 17, 1:9b.

26. See Anna Seidel, *La divinization de Lao Tseu dans le taoïsme des Han* (Paris: Publications de l'École Francaise d'Extrême-Orient, vol. 71, 1969).

27. *LZ* 22, 1:12b.

28. *LZ* 25, 1:13b–14b.

29. *LZ* 25, 13b–14b.

30. See Sarah Allan, *The Heir and the Sage: Dynastic Legend in Early China* (San Francisco: Chinese Materials Center, 1981).

31. *ZZ* 1, 1:5b–6a.

32. *ZZ* 17, 6:14b–15a.

33. *LZ* 16, 1:8b–9a.

34. *LZ* 29, 1:16b–17a.

35. *LZ* 5, 1:3b.

36. *LZ* 52, 2:10a–10b.

37. *LZ* 49, 2:8a.

38. *LZ* 7, 1:4b.

39. *ZZ* 25, 8:23b–24a.

40. See ch. 1, note 2.

41. This combination of Sage and King, however, works on a model very different from the one employed in Confucian readings of the *Laozi* that I discussed in the previous section. Huangdi is a King who becomes a Sage, not a Sage who becomes a King. For more on this theme, see Thomas Michael, "'Huangdi Had 25 Sons': Early Chinese Myths of the First Emperor" (forthcoming).

42. *HDSJ*, 68.

43. *HDSJ*, 72.

44. *HDSJ*, 104.

45. *HDSJ*, 138.

46. *HDSJ*, 110.

47. *HDSJ*, 166–168.

48. *HDSJ*, 142.

49. *XZ* 17, 11:9a–9b.

50. *LZ* 42, 2:5a–5b.

51. See Alan Chan, *Two Visions of the Way* (Albany: State University of New York Press, 1991).

52. Don Harper, "The Bellows Analogy in Laozi V and Warring States Macrobiotic Hygiene," *Early China* 20 (1995): 381–392.

53. *LZ* 5, 1:3b.

54. Early Confucian discourse also speaks of the Dao of antiquity, but it means something very different from the pristine Dao; as Confucius is recorded as saying, "Humans enlarge the Dao, it is not the Dao that enlarges humans."

55. *LZ* 81, 2:24a.

56. *LZ* 77, 2:22b.

57. See Livia Kohn, *The God of the Dao: Lord Lao in History and Myth* (Ann Arbor: University of Michigan Press, 1998).

58. *LZ* 1, 1:1a–1b.

59. *LZ* 6, 1:4a.

60. *LZ* 10, 1:5a–6a.

61. *NY* 13, 16.3b.

62. *LZ* 4, 1:2b–3a.

63. *NY* 8, 16.4a.

64. *NY* 8, 16.4a.

65. For a different view, see Puett, *To Become a God.*

66. *ZZ* 1, 1:4b.

67. *ZZ* 1, 1:5a.

68. *ZZ* 1, 1.6b–7a.

69. *ZZ* 23, 8:7b–8a.

70. *ZZ* 12, 5:5a–5b.

71. *ZZ* 11, 4:18a–19b.

72. *ZZ* 1, 1:9b.

73. *HNZ* 1, 1:1b.

74. *HNZ* 1, 1:2b–3a.

CHAPTER FOUR

1. *LZ* 42, 2:5a.

2. *LZ* 42, 2:5a.

3. *ZZ* 26, 9:5a.

4. *LZ* 23, 1:13a.

5. *LZ* 20, 1:10b–11a.

6. *LZ* 2, 1:1b–2a.

7. *LZ* 12, 1:6b.

8. *LZ* 2, 1:1b.

9. *HNZ* 1, 1:8a.

10. *LZ* 48, 2:7b–8a.

11. *LZ* 18, 1:10a.

12. *LZ* 5, 1:3b.

13. *LZ* 19, 1:10a–10b.

14. *LGDLZ*, 30.

15. *LZ* 38, 2:1a–1b.

16. *LZ* 29, 1:16b–17a.

17. *LZ* 64, 2:17a–17b.

18. *LZ* 58, 2:13b.

19. Other relevant passages include *Laozi* 46, 53, 57, 65, 72, 74, and 75; however, I will not discuss them, as they are more or less fully understandable in the context of the passages already discussed.

20. See, for example, *Essays on Skepticism, Relativism, and Ethics in the Zhuangzi*, eds. Paul Kjellberg and Philip Ivanhoe (Albany: State University of New York SUNY Press, 1996).

21. *ZZ* 2, 1:12b.

22. *ZZ* 2, 1:10a.

23. *ZZ* 2, 1:10a.

24. I expect the distinction that I make between *wu* and *wo* will not go entirely unchallenged on the grounds that *wu* is often taken as a normal subject, and *wo* is regularly used as an object. I want to point out that many concordances of early Chinese texts show *wo* used more as subject than object, at least in terms of English grammar. The early Chinese texts, moreover, demonstrate no conception of the subject/object distinction, as we know it. See also Kuang-ming Wu, *The Butterfly as Companion: Meditations on the First Three Chapters of the Chuang-tzu* (Albany: State University of New York Press, 1990), 155.

25. *ZZ* 2, 1:12a–12b.

26. These and similar passages from the *Qiwulun* are very reminiscent of Martin Heidegger's writings concerning mood and what he calls "being-in-the-world," but I am not familiar with modern studies that pursue this comparison. See Martin Heidegger, *Being and Time*, trs. John Macquarrie and Edward Robinson (New York: Harper and Row, 1962), especially 172–179.

27. *ZZ* 2, 1:12b–13a.

28. *ZZ* 2, 1:13a–14a.

29. *ZZ* 2, 1:14a.

30. *ZZ* 2, 1:14b.

31. In this instance, I emend the *shi* to *wo*. My reasons for doing so are, first, because throughout this chapter *shi* is commonly paired with *fei* in the phrase "assent and rejection," and *wo* is commonly paired with *bi* in the phrase "self and other"; this emendation restores the symmetry of the latter pair, and is entirely in keeping with the style of *Zhuangzi* 2. Second, this emendation allows the wider argument of this section of the chapter to flow much more smoothly. The present argument puts forward ontological claims about the status of knowledge (*bi-wo*, "other and self"), not metaphysical claims about the status of reality (*bi-shi*, "other and this").

32. *ZZ* 2, 1:14b.

33. *ZZ* 2, 1:14b–15a.

34. *ZZ* 2, 1:15a–15b.

35. *ZZ* 2, 1:15b–16b.

36. *LZ* 27, 1:15b.

37. *ZZ* 2, 1:16b–17a.

38. *ZZ* 2, 1:17b.

39. *ZZ* 2, 1:18a.

40. These issues would come to play a central role in later Chan thought. For more on the Chan arguments about sudden and gradual, see Bernard Faure, *The Rhetoric of Immediacy* (Princeton: Princeton University Press, 1991), especially ch. 2.

41. *ZZ* 2, 1:18a–18b.

42. *ZZ* 2, 1:18b–19a.

43. *ZZ* 2, 1:19a.

44. *ZZ* 2, 1:19a.

45. *ZZ* 2, 1:19a–19b.

46. *ZZ* 2, 1:19b.

47. *ZZ* 2, 1:20a–20b.

48. *ZZ* 2, 1:21b.

49. *ZZ* 2, 1:21b–22a.

50. *ZZ* 2, 1:22b–22a.

CHAPTER FIVE

1. *HNZ* 7, 7:6a.

2. Bryan Turner, *Regulating Bodies: Essays in Medical Sociology* (London: Routledge, 1992), 61.

3. Turner, *Regulating Bodies*, 35.

4. Turner, *Regulating Bodies*, 36.

5. As the original text does not provide section breaks, I have adopted the method of numbering the sections of the text according to the standard translation of Allyn Rickett, *Guanzi: Political, Economic, and Philosophical Essays from Early China* (Princeton: Princeton University Press, 1998).

6. For more on the Daoist representation at Jixia, see, for example, Zhang Bingnan, *Jixia Gouchen* (Shanghai: Shanghai Guji Chubanshe, 1991).

7. *NY* 12, 16:5a.

8. *NY* 5, 16:2a.

9. *NY* 4, 16:2a.

10. *NY* 1, 16:1a.

11. *NY* 13, 16:5b.

12. *NY* 10, 16:4b.

13. *NY* 9, 16:4a.

14. *NY* 3, 16:1b.

15. *NY* 11, 16:5a–5b.

16. *NY* 12, 16:5b.

17. *NY* 2, 16:1b.

18. *NY* 4, 16:2a.

19. *NY* 7, 16:3a.

20. *NY* 8, 16:4a.

21. For entirely different reading of the senses in early China, see Jane Geaney, *On the Epistemology of the Senses in Early Chinese Thought* (Honolulu: University of Hawaii Press, 2002).

22. This is Harold Roth's reading; see *Original Tao: Inward Training (Nei-yeh) and the Foundations of Taoist Mysticism* (New York: Columbia University Press, 1999).

23. *NY* 15, 16:6b.

24. *MZ*, 3A.

25. *MZ*, 3A:5a–5b.

26. My comparison of these sections of the *Neiye* and the *Mencius* is in part a response to similar ideas found in Anne Cheng, *Histoire de la pensé chinoise* (Paris: Editions du Seuil, 1997).

27. *MZ*, 3A:4a.

28. *MZ*, 3A:5a.

29. *MZ*, 3A:5a.

30. *MZ*, 3A:6a.

31. Turner, *Regulating Bodies*, 61.

32. Note that "expenditures" 費 (*fei*) is itself a metaphor for physical dissipation, especially in Daoist discourse.

33. *LZ* 44, 1:6a–6b.

34. *LZ* 21, 1:11b–12a.

35. *LZ* 55, 2:11b–12a.

36. *LZ* 4, 1:2b.

37. *LZ* 10, 1:5a–5b.

38. *LZ* 28, 1:16a–16b.

39. Philip Ivanhoe, "The Concept of *de* [Virtue] in the *Laozi*," in *Religious and Philosophical Aspects of the Laozi*, eds. Mark Csikszentmihaly and Philip Ivanhoe (Albany: State University of New York Press, 1999), 249.

40. Ivanhoe, "The Concept of *de* [Virtue] in the *Laozi*," 240.

41. *LZ* 59, 2:14a–14b.

42. *LZ* 52, 2:10a–10b.

43. *LZ* 5, 1:3b.

44. *LZ* 48, 2:7b–8a.

45. *LZ* 12, 1:6b.

46. *LZ* 22, 1:12b.

47. *LZ* 55, 2:11b–12a.

48. *LZ* 76, 2:22a–22b.

49. *ZZ* 22, 7:23a.

50. *ZZ* 5, 2:21a–21b.

51. *ZZ* 23, 8:2a–3a.

52. *ZZ* 32, 10:9a.

53. *ZZ* 32, 10:7a.

54. *ZZ* 31, 10:4a.

55. Parenthetically, the physiological alchemists of later Daoism took up this phrase as a cardinal lesson; further, it was also applied as a major allegorical structuring principle in the Ming dynasty novel, *The Journey to the West* 西游記 (*Xiyouji*).

56. *ZZ* 3, 2:1a–1b.

57. *ZZ* 4, 2:4b–6a.

58. *ZZ* 4, 2:7a–8a.

59. *Dhammapada*, trs. John Ross Carter and Mahinda Palihawadana (Oxford: Oxford University Press, 1987), 14.

60. A. C. Graham, *Chuang-tzu* (London: Mandala, 1991), 23–24.

61. *ZZ* 6, 3:5b.

62. *ZZ* 6, 3:14a–14b.

63. See, for example, *LY* 6.3, 11.7.

64. *ZZ* 11, 4:14b.

65. *ZZ* 5, 2:23a–23b.

66. For a detailed discussion of *qing* in Chinese discourse, see Anthony Yu, *Rereading the Stone* (Princeton: Princeton University Press, 1997), especially ch. 2.

67. Burton Watson, *Chuang Tzu: Basic Writings* (New York: Columbia University Press, 1996), 72; Victor Mair, *Wandering on the Way: Early Taoist Tales and Parables of Chuang Tzu* (Hawaii: University of Hawaii Press, 1994), 49; Graham, *Chuang-Tzu* (1991), 82.

68. *ZZ* 31, 10:5b–6a.

69. *ZZ* 6, 3:1a–2a.

70. *ZZ* 5, 2:22b–23a.

71. A different argument that follows a different reading of the closing of the gap between Heaven and humans can be found in Chad Hansen, *A Daoist Theory of Chinese Thought* (Oxford: Oxford University Press, 1992), especially ch. 5.

72. *ZZ* 15, 6:2b–3a.

73. *ZZ* 19, 7:1a–1b.

74. Charles Le Blanc, *Huai Nan Tzu: Philosophical Synthesis in Early Han Thought* (Hong Kong: Hong Kong University Press, 1985), 83.

75. *HNZ* 1, 1:4a.

76. *HNZ* 1, 1:6b.

77. *HNZ* 1, 1:12a–12b.

78. Harold Roth, "Psychology and Self-Cultivation in Early Taoistic Thought," *Harvard Journal of Asiatic Studies* 51 (1991): 599–650.

79. *HNZ* 1, 1:16a–16b.

80. *HNZ* 2, 2:8a–8b.

81. *HNZ* 2, 2:9b–10b.

82. *HNZ* 2, 2:11b–12a.

83. *HNZ* 7, 7:12a–12b.

84. *HNZ* 7, 7:1a–1b.

85. For a detailed study of this, see Aihe Wang, *Cosmology and Political Culture in Early China* (Cambridge: Cambridge University Press, 2000).

86. For more on this text, see Paul Unschuld, *Huang Di Nei Jing Su Wen* (Berkeley: University of California Press, 2003).

87. *HNZ* 7, 7:1.b.

88. *HNZ* 7, 7:1b–2a.

89. *HNZ* 7, 7:2a.

90. *HNZ* 7, 7:2b.

91. *HNZ* 7, 7:2b–3a.

92. *HNZ* 7, 7:6b.

93. *HNZ* 7, 7:5a–6a.

CHAPTER SIX

1. For a translation of the CCP state document laying this out, see Donald MacInnis, *Religion in China Today* (Maryknoll, NY: Orbis Books, 1989), 8–26.

2. See Plato, *The Symposium and the Phaedrus: Plato's Erotic Dialogues*, tr. William Cobb (Albany: State University of New York Press, 1993).

3. Pierre Hadot, *Philosophy as a Way of Life*, tr. Arnold Davidson (Cambridge: Blackwell Publishers, 1995).

4. See Jean-Paul Sartre, *Being and Nothingness*, tr. Hazel Barnes (New York: Citadel Press, 1965).

5. See Martin Heidegger, *Being and Time*, tr. John Macquarrie and Edward Robinson (New York: Harper and Collins, 1962).

6. Brian Smith, *Classifying the Universe* (Oxford: Oxford University Press, 1997).

7. Jeffrey Broughton, *The Bodhidharma Anthology* (Berkeley: University of California Press, 1999), 21.

Bibliography

PRIMARY SOURCES

DY: Daoyuan 道原. Critical text from Robin Yates, *Five Lost Classics*. New York: Ballantine Books, 1997. Middle third century BCE.

GDLZ: Guodian Laozi 郭店老子. Critical text from Robert Henricks, *Lao Tzu's Tao Te Ching*. NewYork: Columbia University Press, 2000. Late fourth century BCE.

HDSJ: Huangdi Sijing 黃帝四經. Critical text from Robin Yates, *Five Lost Classics*. New York: Ballantine Books, 1997. Middle third century.

HNZ: Huainanzi 淮南子. *Sibu Beiyao.* 139 BCE.

LiZ: Liezi 列子. *Sanmin Shuju*. Second century CE.

LSCQ: Lüshi Chunqiu 呂氏春秋. *Sibu Beiyao.* Middle third century BCE.

LY: Lunyu 論語正義. *Sibu Beiyao.* Middle fourth century BCE.

LZ: Laozi 老子. *Sibu Beiyao.* Third century BCE.

MWDLZ: Mawangdui Laozi 馬王堆老子. Critical text from D. C. Lau, *Tao Te Ching*. Hong Kong: Chinese University Press, 1989. Late third century BCE.

MZ: Mengzi 孟子. *Sibu Beiyao.* Late fourth century BCE.

NY: Neiye 內業, in *Guanzi* 管子 49. *Sibu Beiyao.* Middle fourth century BCE.

SD: Shui Di 水地, in *Guanzi* 管子 39. *Sibu Beiyao.* Middle second century BCE.

TYSS: Taiyi Sheng Shui 太一生水. Critical text from Robert Henricks, *Lao Tzu's Tao Te Ching*. New York: Columbia University Press, 2000. Late fourth century BCE.

XZ: Xunxi 荀子. *Sibu Beiyao.* Middle third century BCE.

ZZ: Zhuangzi 莊子. *Sibu Beiyao.* Middle third century BCE.

MODERN TEXTS

Allan, Sarah. *The Heir and the Sage: Dynastic Legend in Early China.* San Francisco: Chinese Materials Center, 1981.

Allan, Sarah. *The Way of Water and Sprouts of Virtue.* Albany: State University of New York Press, 1997.

Allan, Sarah, and Crispin Williams, eds. *The Guodian Laozi: Proceedings of the International Conference, Dartmouth College, May 1998.* Berkeley, Society for the Study of Early China and Institute of East Asian Studies, University of California, 2000.

Ames, Roger. "Introduction," in *Wandering at Ease in the Zhuangzi*, ed. Roger Ames. Albany: State University of New York Press, 1998.

Boltz, William. "Kung Kung and the Flood: Reverse Euhemerization in the *Yao Tian*," *T'oung Pao* 67, 1981.

Broughton, Jeffrey. *The Bodhidharma Anthology: The Earliest Records of Zen.* Berkeley: University of California Press, 1999.

Chan, Alan. *Two Visions of the Way.* Albany: State University of New York Press, 1991.

Cheng, Anne. *Histoire de la pensé chinoise.* Paris: Editions du Seuil, 1997.

Creel, Herlee. *What Is Taoism? And Other Studies in Chinese Cultural History.* Chicago: University of Chicago Press, 1970.

Dhammapada, trs. John Ross Carter and Mahinda Palihawadana. Oxford: Oxford University Press, 1987.

Eliade, Mircea. *The Myth of the Eternal Return: Or, Cosmos and History*, tr. Willard Trask. Princeton: Princeton University Press, 1991.

Faure, Bernard. *The Rhetoric of Immediacy: A Cultural Critique of Chan/Zen Buddhism.* Princeton: Princeton University Press, 1991.

Geaney, Jane. *On the Epistemology of the Senses in Early Chinese Thought.* Honolulu: University of Hawaii Press, 2002.

Girardot, Norman. *Myth and Meaning in Early Taoism: The Theme of Chaos (Hun-Tun).* Berkeley: University of California Press, 1983.

Graham, A.C. tr. *Chuang-tzu: The Inner Chapters: A Classic of Tao.* London: Mandala, 1991.

Graham, A.C. *A Companion to Angus C. Graham's Chuang Tzu*, ed. Harold Roth. Hawaii: University of Hawaii Press, 2003.

Hadot, Pierre. *Philosophy as a Way of Life*, tr. Arnold Davidson. Cambridge: Blackwell Publishers, 1995.

Hansen, Chad. *A Daoist Theory of Chinese Thought: A Philosophical Interpretation.* Oxford: Oxford University Press, 1992.

Harper, Don. "The Bellows Analogy in *Laozi* V and Warring States Macrobiotic Hygiene," in *Early China* 20 (1995): 381–392.

Harper, Don. "The Nature of *Taiyi* in the Guodian Manuscript '*Taiyi sheng shui*': Abstract Cosmic Principle or Supreme Cosmic Deity?" Paper prepared for the "International Symposium on the Guodian Chu Bamboo Slips and Related Excavated Materials." December, 2000.

Heidegger, Martin. *Being and Time*, tr. John Macquarrie and Edward Robinson. New York: Harper and Collins, 1962.

Husserl, Edmund. *The Crisis of European Sciences and Transcendental Phenomenology: An Introduction to Phenomenological Philosophy.* Evanston: Northwestern University Press, 1970.

Ivanhoe, Philip. "The Concept of *de* [Virtue] in the *Laozi*," in *Religious and Philosophical Aspects of the Laozi*, ed. Mark Csikszentmihaly and Philip Ivanhoe. Albany: State University of New York Press, 1999.

Kjellberg, Paul, and Philip Ivanhoe, eds. *Essays on Skepticism, Relativism, and Ethics in the Zhuangzi*. Albany: State University of New York Press, 1996.

Kohn, Livia. *The God of the Dao: Lord Lao in History and Myth*. Ann Arbor: University of Michigan Press, 1998.

Lagerwey, John. *Taoist Ritual in Chinese Society and History*. New York: Macmillan, 1987.

Lakoff, George, and Mark Johnson. *Philosophy in the Flesh: The Embodied Mind and Its Challenge to Western Thought*. New York: Basic Books, 1999.

Le Blanc, Charles. *Huai Nan Tzu: Philosophical Synthesis in Early Han Thought*. Hong Kong: Hong Kong University Press, 1985.

Lewis, Mark. *Sanctioned Violence in Early China*. Albany: State University of New York Press, 1990.

Li, Ling. "An Archaeological Study of Taiyi 太一 (Grand One) Worship," tr. Don Harper. *Early Medieval China* 2 (1995–1996): 1–39.

Li Ling and Constance Cook, "Appendix: Translation of the Chu Silk Manuscript," in *Defining Chu: Image and Reality in Ancient China*, ed. Constance Cook and John Major. Hawaii: University of Hawaii Press, 1999.

Liu Xiaogan, *Classifying the Zhuangzi Chapters*. Ann Arbor: University of Michigan Press, 1994.

MacInnis, Donald. *Religion in China Today*. Maryknoll, NY: Orbis Books, 1989.

Mair, Victor. *Wandering on the Way: Early Taoist Tales and Parables of Chuang Tzu*. Hawaii: University of Hawaii Press, 1994.

Major, John. "Characteristics of Late Chu Religion," in *Defining Chu: Image and Reality in Ancient China*, ed. Constance Cook and John Major. Honolulu: University of Hawaii Press, 1999.

Mathieu, Rémi. *Étude sur la mythologie et l'ethnologie de la Chine ancienne*. Paris: Institut des Hautes Études chinoises, 1983.

Michael, Thomas. "Debating the Spirit in Early China" (review article), in *Journal of Religion*, v. 83 (2003) 421–429.

Michael, Thomas. "'Huangdi Had 25 Sons': Early Chinese Myths of the First Emperor" (forthcoming).

Michael, Thomas. "Confucius and Laozi: Two Faces of the Dao," in *Metaphilosophy and Chinese Thought: Interpreting David Hall*, eds. Ewing Chinn and Henry Rosemont, Jr. New York: Global Scholarly Publications, 2005.

Plato. *The Symposium and the Phaedrus: Plato's Erotic Dialogues*, tr. William Cobb. Albany: State University Press, 1993.

Peerenboom, R. P. *Law and Morality in Ancient China: The Silk Manuscripts of Huang-Lao*. Albany: State University of New York Press, 1993.

Porter, Deborah. *From Deluge to Discourse: Myth, History, and the Generation of Chinese Fiction*. Albany: State University of New York Press, 1996.

Puett, Michael. *To Become A God: Cosmology, Sacrifice, and Self- Divinization in Early China*. Cambridge: Harvard University Press, 2002.

Rickett, Allyn. *Guanzi: Political, Economic, and Philosophical Essays from Early China*, Volume 1 and 2. Princeton: Princeton University Press, 1985, 1998.

Roetz, Heiner. *Confucian Ethics of the Axial Age: A Reconstruction Under the Aspect of the Breakthrough Toward Post-conventional Thinking.* Albany: State University of New York Press, 1993.

Roth, Harold. "Psychology and Self-Cultivation in Early Taoistic Thought." *Harvard Journal of Asiatic Studies* 51 (1991): 599–650.

Roth, Harold. *Original Tao: Inward Training [Nei-yeh] and the Foundations of Taoist Mysticism*. New York: Columbia University Press, 1999.

Sartre, Jean-Paul. *Being and Nothingness: An Essay in Phenomenological Analysis*, tr. Hazel Barnes. New York: Citadel Press, 1965.

Schipper, Kristofer. *The Taoist Body*, tr. Karen Duval. Berkeley: University of California Press, 1993.

Seidel, Anna. *La divinisation de Lao tseu dans le taoïsme des Han*. Paris: L'École française d'Extrême-Orient, 1969.

Sivin, Nathan. "The Myth of the Naturalists," in *Medicine, Philosophy and Religion in Ancient China*. Great Britain: Variorum, 1995.

Sivin, Nathan, and Geoffrey Lloyd. *The Way and the Word: Science and Medicine in Early China and Greece.* New Haven: Yale University Press, 2002.

Smith, Brian K. *Classifying the Universe: The Ancient Indian Varna System and the Origins of Caste.* Oxford: Oxford University Press, 1997.

So, Jenny. "Chu Art: Link Between the Old and New," in *Defining Chu: Image and Reality in Ancient China*, ed. Constance Cook and John Major, 33–47. Honolulu: University of Hawaii Press, 1999.

Swanson, Gerald. *The Great Treatise: Commentary Tradition to the "Book of Changes."* Ann Arbor, Michigan: UMI Dissertation Services, 1974.

Turner, Bryan. *Regulating Bodies: Essays in Medical Sociology.* London: Routledge, 1992.

Unschuld, Paul. *Huang Di Nei Jing Su Wen: Nature, Knowledge, Imagery in an Ancient Chinese Medical Text*. Berkeley: University of California Press, 2003.

Wang, Aihe. *Cosmology and Political Culture in Early China*. Cambridge: Cambridge University Press, 2000.

Watson, Burton. *Chuang Tzu: Basic Writings.* New York: Columbia University Press, 1996.

Wu, Kuang-ming. *The Butterfly as Companion: Meditations on the First Three Chapters of the Chuang-tzu*. Albany: State University of New York Press, 1990.

Yu, Anthony. *Rereading the Stone: Desire and the Making of Fiction in Dream of the Red Chamber.* Princeton: Princeton University Press, 1997.

Zhang Bingnan, *Jixia Gouchen*. Shanghai: Shanghai Guji Chubanshe, 1991.

Index